Machine Learning Methods for Signal, Image and Speech Processing

RIVER PUBLISHERS SERIES IN SIGNAL, IMAGE AND SPEECH PROCESSING

Series Editors:

MARCELO SAMPAIO DE ALENCAR
Universidade Federal da Bahia UFBA, Brasil

MONCEF GABBOUJ
Tampere University of Technology, Finland

THANOS STOURAITIS
University of Patras, Greece
and
Khalifa University, UAE

The "River Publishers Series in Signal, Image and Speech Processing" is a series of comprehensive academic and professional books which focus on all aspects of the theory and practice of signal processing. Books published in the series include research monographs, edited volumes, handbooks and textbooks. The books provide professionals, researchers, educators, and advanced students in the field with an invaluable insight into the latest research and developments.

Topics covered in the series include, but are by no means restricted to the following:

- Signal Processing Systems
- Digital Signal Processing
- Image Processing
- Signal Theory
- Stochastic Processes
- Detection and Estimation
- Pattern Recognition
- Optical Signal Processing
- Multi-dimensional Signal Processing
- Communication Signal Processing
- Biomedical Signal Processing
- Acoustic and Vibration Signal Processing
- Data Processing
- Remote Sensing
- Signal Processing Technology
- Speech Processing
- Radar Signal Processing

For a list of other books in this series, visit www.riverpublishers.com

Machine Learning Methods for Signal, Image and Speech Processing

Editors

M. A. Jabbar
Vardhaman College of Engineering, India

Kantipudi MVV Prasad
Symbiosis Institute of Technology, India

Sheng-Lung Peng
National Dong Hwa University, Taiwan

Mamun Bin Ibne Reaz
Universiti Kebangsaan, Malaysia

Ana Madureira
Interdisciplinary Studies Research Center, Portugal

Published, sold and distributed by:
River Publishers
Alsbjergvej 10
9260 Gistrup
Denmark

www.riverpublishers.com

ISBN: 978-87-7022-369-0 (Hardback)
      978-87-7022-368-3 (Ebook)

©2021 River Publishers

Contents

Preface

This book describes in detail the applications of machine learning (ML) in signal, image, and speech processing. Important research findings and recent innovations in the field of ML and signal processing (SP) are presented in this book. This book will provide valuable information to both researchers and practitioners with the latest advances and trends in ML and SP.

The SP landscape has been enriched by recent advances in artificial intelligence (AI) and ML, especially since 2011 or so, yielding new tools for signal estimation, classification, prediction, and manipulation. Layered signal representations, nonlinear function approximation, and nonlinear signal prediction are now feasible at large scale in both dimensionality and data size. These are leading to significant performance gains in a variety of long-standing problem domains like speech and image.

This book will help academicians, researchers, developers, graduate, and undergraduates to comprehend the potential in exploring social multimedia signal data collected from social media network from a SP perspective. Chapters in this book focus on AI utilization in the speech, image, communications, and virtual reality domains.

The content of this book is a distillation of many chapters which have subsequently become the material for applications of ML in signal, image, and speech processing.

This edited book covers various applications and techniques using ML in image, signal, and speech processing.

Chapter 1 presents the evaluation of adaptive algorithms for the recognition of cavities in industry. This article anticipates to enterprise unique cavities recognition scheme that comprises three segments: (a) enhancement and noise removing, (b) feature abstraction, and (c) cataloguing.

Chapter 2 presents lung cancer prediction using feature selection and recurrent residual convolutional neural network (RRCNN). The proposed model makes use of three underlying algorithms, that is, UNet, residual network, as well as recurrent convolutional neural network (RCNN).

Chapter 3 presents ML application for detecting leaf diseases with image processing schemes. This chapter reports on the evolution and performance of ML with image processing in the detection the plant leaf diseases.

In Chapter 4, the authors present COVID-19 forecasting using deep learning models. In this chapter, forecast models comprising various AI approaches, such as support vector regression (SVR), long- and short-term memory (LSTM), and bidirectional long- and short-term memory (Bi-LSTM) are assessed for time series prediction of confirmed cases, deaths, and recoveries in 10 major countries affected with COVID-19.

Chapter 5 discusses the 3D smart learning using ML technique. The proposed system is designed to learn the correspondence between preached words and conceptual visual attributes from a spoken image description dataset.

Chapter 6 discusses the SP for OFDM spectrum sensing approaches in cognitive networks. This chapter presents the enhanced and comparative analysis of OFDM spectrum sensing approaches in cognitive networks.

Chapter 7 highlights an ML algorithm for biomedical SP application. In this chapter, significant emphasis has been given to ML for biomedical SP. The basic purpose of this chapter is to explore the numerous possibilities of ML in the field of biomedical SP.

Chapter 8 presents reversible image data hiding based on prediction-error of prediction-error histogram (PPEH). In this chapter, the authors proposed an unique reversible data hiding technique based on PPEH of an element to reversibly carry the key data.

Chapter 9 comes with a concept object detection using deep convolution neural network. In this chapter, the authors present a survey on object detection algorithms. Various datasets, such as PASCAL VOC versions, ILSVRC, MS COCO, and Open of object detection (OICOD) information has also been provided.

In Chapter 10, the authors presented an intelligent patient health monitoring system based on a multi-scale convolutional neural network (MCCN) and Raspberry Pi. In this chapter, significant emphasis has been given to ML for an efficient real-time monitoring system. The basic purpose of this chapter is to explore the numerous possibilities of ML in the field of real-time monitoring systems.

At last, we wish to express our heartful thanks to the authors who contributed the chapters to this edited book, reviewers for reviewing the manuscript, and the River Publisher for accepting this book proposal to publish the book.

Editors:

Dr. M. A. Jabbar
Dr. Kantipudi MVV Prasad
Dr. Sheng-Lung Peng
Dr. Mamun Bin Ibne Reaz
Dr. Ana Madureira

List of Figures

List of Tables

List of Contributors

Ahmad, Syed Jalal, *Malla Reddy Engineering College (main campus), Secunderabad; E-mail: jalalkashmire@gmail.com*

Aleem, Md., *Bhaskar Engineering College, Moinabad; E-mail: Mdaleem80@gmail.com*

Bhargava Choubey, Shruti, *Department of Electronics Communication, Sreenidhi Institute of science and technology, Hyderabad, India; E-mail: shrutibhargava@sreenidhi.edu.in*

Cheguru, Sahiti, *Gokaraju Rangaraju Institute of Engineering and Technology, India; E-mail: sahiticheguru2000@gmail.com*

Choubey, Abhishek, *Department of Electronics Communication, Sreenidhi Institute of science and technology Hyderabad, India; E-mail: abhishek@sreenidhi.edu.in*

Fathima, Syed Aley, *Vardhaman College of Engineering, India; E-mail: noorainmustafa786@gmail.com*

Ghosh, Siddhartha, *Vidya Jyothi Institute of Technology, India; E-mail: siddhartha@vjit.ac.in*

Harini, S., *Vardhaman College of Engineering, India; E-mail: sayiniharini2k@gmail.com*

Jabbar, M. A., *Vardhaman College of Engineering, India; E-mail: jabbar.meerja@gmail.com*

Krishna Prasad, P.E.S.N., *CSE, GIT, GITAM Deemed University, Visakhapatnam-530045, India; E-mail: surya125@gmail.com*

Kurumbanshi, Suresh, *SVKMs NMIMS, MPSTME, India; E-mail: sureshkurumbanshi@gmail.com*

Nagajyothi, D., *Vardhaman College of Engineering, India; E-mail: d.nagajyothi@vardhaman.org*

Nirkhi, Smita, *GHRIET, Nagpur, India; E-mail: smita811@gmail.com*

Padmavathi, V, *Anurag Univerity, India;*
E-mail: chpadmareddy1@gmail.com

Patil, Shashikant, *SVKMs NMIMS, MPSTME, India;*
E-mail: sspatil999@gmail.com

Prerana, C.H., *Vidya Jyothi Institute of Technology, India;*
E-mail: preranacheguru@gmail.com

Saba Raoof, Syed, *VIT University, India;*
E-mail: syedsabaraoof@gmail.com

Satish Kumar, G.A.E., *Vardhaman College of Engineering,*
Hyderabad, India; E-mail: gaesathi@gmail.com

Sonawane, Sachin, *SVKMs NMIMS, MPSTME, India;*
E-mail: sachin.sonawane@nmims.edu

Srilatha, M., *Vardhaman College of Engineering, India;*
E-mail: m.srilatha@vardhaman.org

Srinivasulu Reddy, D., *ECE, Sri Venkateswara College of Engineering,*
Tirupati-517507, India; E-mail: cnudega@gmail.com

Subba Rao, S.P.V., *Department of Electronics Communication, Sreenidhi*
Institute of science and technology, Hyderabad, India;
E-mail: spvsubbarao@sreenidhi.edu.in

Sujatha, Canavoy Narahari, *Sreenidhi Institute of Science and Technology,*
India; E-mail: cnsujatha@sreenidhi.edu.in

Sumalatha, R., *Vardhaman College of Engineering, Hyderabad; India;*
E-mail: amrutha.suma18@gmail.com

Sushanth, T., *Vardhaman College of Engineering, India;*
E-mail: thadishetti.sushanth@gmail.com

Tarine, Deepthi, *Vidya Jyothi Institute of Technology, India;*
E-mail: tarinedeepthi1998@gmail.com

Tejasree, K., *Vidya Jyothi Institute of Technology, India;*
E-mail: tejasreekomati05@gmail.com

Unissa, Ishrath, *CMR Technical Campus, Secunderabad;*
E-mail: ishrathunnisa94@gmail.com

Upendra Raju, K., *ECE, Sri Venkateswara College of Engineering, Tirupati-517507, India; E-mail: kupendraraju@gmail.com*

Vijayalata, Y., *Gokaraju Rangaraju Institute of Engineering and Technology, India; E-mail: vijaya@griet.ac.in*

List of Abbreviations

1D	One Dimension
2D	Two Dimension
3D	Three Dimension
AI	Artificial Intelligent
ANN	Artificial Neural Networks
AUC	Area Under Curve
CityGML	City Geographic Markup Language
CNN	Convolutional neural network
CNN	Convolutional Neural Networks
CycleGAN	Cycle Generative Adversarial Network
DA	Dragonfly Algorithm
DAQ	Data Acquisition
DBSCAN	Density-Based Spatial Clustering of Applications with Noise
DCNN	Deep Convolutional neural network
DCNN	Deep Convolutional NN
DMFT	Decayed, Missing and Filled Teeth
DT	Decision Trees
ETIAS	European Travel Information and Authorization System
FDR	False Discovery Rate
FNR	False Negative Rate
FOR	False Omission Rate
FPR	False Positive Rate
GAN	Generative Adversarial Network
HOG	Histogram of Oriented Gradients
HOG	Histogram of Gradients
ILSVRC	ImageNet Large Scale Visual Recognition Challenge
KNN	k-nearest neighbors
KNN	K- Nearest Neighbor
LabVIEW	Laboratory Virtual Instrument Engineering Workbench
LBP	Local binary pattern

LiDAR	Light Detection and Ranging
LM	Levenberg-Marquardt
MCC	Matthews Correlation Coefficient
ML	Machine Learning
MLR	Multiple Logistic Regressions
MNP-ADA	MPCA model Nonlinear Programming with Adaptive DA
MS COCO	Microsoft Common Objects in COntext
NB	Naive Bayes
NI	National Instruments
NILTI	Near-Infrared Light Trans Illumination
NN	Neural Network
NNE	Neural Network Ensemble
NPV	Net Present Value
OCR	Optical Character Recognition
OICOD	Open of object detection
PSO	Particle Swarm Optimization
RBF	Radial Basis Functions
RCNN	Regions with CNN
R-CNN	Region based Convolutional Neural Networks
ReLu	Rectified Linear Unit
RF	Random Forest
RGB	Red, Green and Blue
RMSE	Root Mean Square Error
RNA	Ribonucleic acid
ROI	Region of Interest
SIDF	Side-Inner Difference Features
SK-learn	Scikit-learn
SSD	Single Shot MultiBox Detector
SSF	Symmetrical Similarity Features
SURF	Speeded-up Robust Features
SVM	Support Vector Machine
SVM	Support Vector Machine
SVM	Support Vector Machines
USB	Universal Serial Bus
VC	Voice Conversion
VOC	Visual Object Classes
YOLO	You Only Look Once

1

Evaluation of Adaptive Algorithms for Recognition of Cavities in Dentistry

Shashikant Patil[1], Smita Nirkhi[2], Suresh Kurumbanshi[3], and Sachin Sonawane[4]

[1,3,4]SVKMs NMIMS, MPSTME, India
[2]GHRIET, Nagpur, India
E-mail: sspatil@ieee.org; smita811@gmail.com; sureshkurumbanshi@gmail.com; sachin.sonawane@nmims.edu

Abstract

The former exploration of oral cavities in dentistry has accomplished abundant attention as it clues to carious lacerations. The prevailing cavities recognition procedures have been unsuccessful to diagnose the enamel lacerations at the prior phase. This article anticipates to enterprise a unique cavities recognition scheme that comprises three segments: (a) Enhancement and Noise removing (b) Feature Abstraction (c) Cataloguing. In the preliminary segment, the superiority of the image is enhanced by some of the signal and image enriching approaches such as disparity upgrading, grey thresholding, and active contour, correspondingly. Consequently, the topographies get extracted by means of Multi-Linear Principal Component Analysis (MPCA) and Multi-Linear Linear Discriminant Analysis (MLDA). Finally, cataloging is done using Artificial Neural Network (ANN), which is taught by the adaptive Dragonfly algorithm (DA) algorithm. The espoused MSL model Nonlinear Programming with Adaptive DA (MNP-ADA) bargains the exact categorized outcomes.

Keywords: Oral Cavities, Image Processing, Cataloguing, CLAHE, MPCA, MLDA, Classifiers.

1

1.1 Introduction

'Dental cavities are a syndrome, demarcated as the procedure of continual demineralization of the inorganic constituent of the teeth convoyed by the crumbling of the organic portion. It can also be demarcated as a vigorous infection progression, in which initial lacerations experience numerous demineralization and remineralization sequences afore being articulated by dentist' [26–28].

As a result, it is necessary to identify the symptoms of caries at an earlier stage rather than searching for cavities [1, 2, 8]. The precise treatment proceeding to cavities would permit besieged precautionary diagnoses like fissure, pits, and fluoride sealants, thus substantially enriching oral healthiness. This curtails the prerequisite for extensive filling and piercing [3–5].

A clinician needs expertise, knack, and understanding to feat the precise pinpointing procedure and to handle them. Filmic and pictorial assessment by means of radiographic films, mouth mirrors, and conformist probes were the investigative practices that existed customarily and used in erstwhile [6, 7, 25].

The perceptible distinctions in the interior tooth conformation distress the micro porosity of coating that successively disturbs the diffusion of light via the coating [24]. The consequences of plentiful readings argue that the deployment of the probe has constrained assessment in perceiving cavities and is also initiate to intrude remineralization [10, 11, 19]. Detecting cavities happens to be a thought-provoking duty for dentists and experts [24–26]. Besides, the programmed identification of cavities is done through intelligent and automated algorithms [9, 12].

The foremost contribution of the article is as below:

At the preliminary stage, the superiority of the image is enhanced by pre-processing procedures like CLAHE, contrast enhancement, grey thresholding, and active contour.

Furthermore, the features are taken out by means of MPCA and MLDA and subsequently classification is carried out using optimized NN, where the artificial DA model is projected for the training procedure.

Lastly, Interpretative exploration is accomplished to authenticate the performance of the suggested prototypical procedure.

The article is organized as shown: Section 1.1 illustrates the scrutiny on oral cavities recognition. Section 1.3 outlines the embraced cavities recognition model and Section 1.4 portrays the reckoning of distance measure. Additionally, Section 1.5 particularises the anticipated artificial dragonfly algorithm for resolving the optimization glitches. Sections 1.6 and 1.7 depicts the conclusions and inference, works appraisal.

1.2 Related Work

In 2019, Ayşe et al. [1] have obtainable a vivo study for confirming the recognition of proximal caries by means of NILTI. Moreover, the diagnostic performance of the device was compared over other caries recognition techniques, together with visual assessment. Accordingly, here a total of nine seventy-four proximal surfaces of stable posterior teeth from thirty-four patients were taken into account. The data were examined with statistical analysis and the AUC, specificity, and sensitivity were computed.

In 2019, Darshan et al. [2] have computed the relationship among susceptibility of dental caries progression risk and ENAM gene polymorphisms. The implemented analysis was performed on one sixty-eight children from South India and kids affected by dental caries were also taken into account. 'Preliminary Insilco analysis' has revealed that variations in 'rs7671281 (Ile648Thr) amino acid' leads to the functional and structural changes in the ENAM.

In 2018, Lee et al. [26] have adopted a method for evaluating the efficiency of DCNN approaches for diagnosis and detection of dental caries on 'periapical radiographs'. Accordingly, this analysis focused on the potential effectiveness of the DCNN framework for the diagnosis and detection of dental caries. From the analysis, the DCNN framework has offered significant performance in recognizing dental caries in 'periapical radiographs'.

In 2019, Yue et al. [27] have carried out an analysis on detecting dental caries on three eighty-six kids residing in Mexico town. Here, 'graphite-furnace atomic-absorption spectroscopy' was used for quantifying the Pb levels of blood. Accordingly, the existence of dental caries was computed by means of DMFT scores. Furthermore, the residual approach was exploited in this work for determining the total energy produced in the children based on the consumption of sweets and beverages.

In 2019, Cácia et al. [28] have analyzed how the risk factors of patients influenced operative diagnostic decisions in a dental oriented system in the Netherlands. In this work, the data were gathered from eleven dental practices and the patients attended the practice regularly throughout the observation time. Consequently, a descriptive study was carried out after performing the MLR process.

1.3 Proposed Model for Cavities Detection

Figure 1.1 reveals the schematic depiction of the embraced dental cavities detection model. The instigated outline comprises three foremost steps:

- Enhancement and Pre-processing;
- Feature Extraction;
- Classification;
- Optimization.

At the outset, the input image Im is imperiled to noise removing, brightening, and enriching through pre-processing, which comprises four important image upgrading features such as CLLAHE, contrast upgrading, grey thresholding, and active contour. From the pre-processed image I_{pre}, the features are mined by the aid of the MSL method like MLDA & MPCA model. These mined features are then imperiled to cataloging using NN classifier that bids the categorized outcome (Cavities or No Cavities) [13–16].

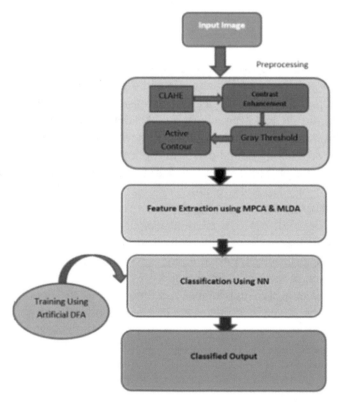

Figure 1.1   Graphical representation of the adopted scheme.

1.3.1 Pre-processing

The image Im is improved by carrying out the below processes.

Conventional Adaptive Histogram Equalisation is apt to over intensify the contrast in near-constant provinces of the image, meanwhile the histogram in such areas is exceedingly strenuous. As a consequence, Adaptive Histogram Equalization may root noise to be enlarged in near-constant areas. Contrast Limited AHE (CLAHE) is modified of adaptable and adjustable histogram equalization in which the dissimilarity intensification is inadequate, so as to diminish this delinquent of noise intensification.

In Contrast Limited AHE (CLAHE), the contrast solidification in the vicinity of a quantified pixel worth is quantified by the gradient of the variation function. This is interactive to the slope of the locality accumulative dissemination function and accordingly to the cost of the histogram at that pixel cost. Contrast Limited AHE confines the intensification by trimming the histogram at a predefined value before calculating the CDF. This confines the slant of the CDF and consequently of the alteration function. The cost at which the histogram is cropped, the ostensible clip perimeter, be governed by normalization of the histogram and thus on the extent of the vicinity region. Collective values limit the resultant intensification. It is advantageous not to discard the part of the histogram that exceeds the clip limit but to redistribute it equally among all histogram bins (refer Figure 1.2) [17–21].

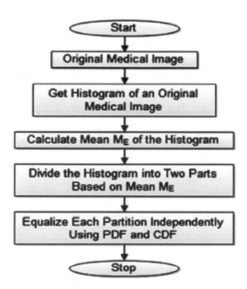

Figure 1.2 Histogram Equalization.

The relocation will impulsion roughly silos over the clip limit again (region shaded green in the figure), resultant in an active clip limit that is more than the prescribed limit and the particular value of which be contingent on the image. If this is disagreeable, the reorganization technique can be recurrent recursively until the excess is negligible.

1.3.2 Contrast Enhancement

The contrasting of the resized input image Im^g is enhanced here. The particular procedure controls the image intensity [16, 22, 23] and thus the image resolution is developed via the brightness and darkness of Im^g, as given by Equation (1.1), in which V refers to the contrast improvement of the image. Therefore, the current Im^g transforms into a grey image Im_{new}^g.

$$V = \begin{pmatrix} (((\text{Im} - low\_in) / (high\_in - low\_in))^{\wedge} gamma) \\ * (high\_out - low\_out) \end{pmatrix} + low\_out$$

(1.1)

Grey thresholding: The Otsu's oriented grey thresholding [20] method portrays the threshold of the image, which is exploited for converting the grey pixel to either black or white. This is performed depending on the grey intensity (refer Figure 1.3).

Active contour [19]: Here, 2 types of driven forces namely, external and internal energy are exploited. This framework gets smoothed via internal forces and it is reallocated in the direction through the external energy. Therefore, the contour $G(n)$ is formed by the coordinate sets such as $l(n)$ and $k(n)$ as given in Equation (1.2), where (k, l) indicates the contour coordinates and denotes the normalized index of the control point.

$$G(n) = (k(n)l(n)); \; G(n) \in \text{Im}_{new}^C(k, l)$$

(1.2)

Equation (1.3) shows the total energy of deformed design, where $\text{Im}^{g^{int}}$ indicates the internal energy of the curve, $\text{Im}^{g^{con}}$ denotes the exterior restriction, denotes the energy of the image.

$$FO^* = \int_0^1 (FO^{\text{int}\,l}G(n) + FO^{im}G(n) + FO^{con}G(n))dn$$

(1.3)

In addition, the bending energy and elastic energy are summed up to form the internal energy as specified in Equation (1.4), where $\alpha(n), \beta(n)$ indicates the

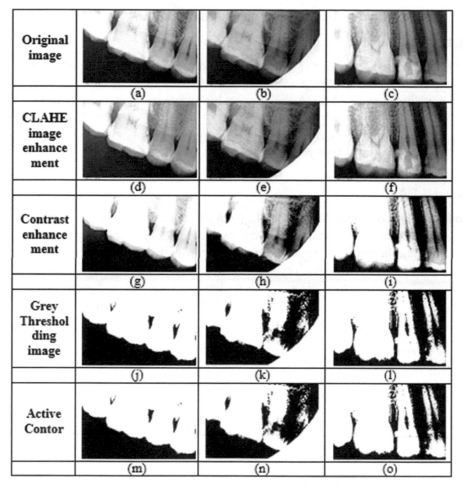

Figure 1.3   (a) – (o) Image Processing of Dental Images.

varying parameter that denotes continuity and contour curving respectively.

$$FO^{\text{int } l} = FO^{elastic} + FO^{bend} = \alpha(n)\left|\frac{du}{dn}\right|^2 + \beta(n)\left|\frac{d^2u}{dn^2}\right|^2 \quad (1.4)$$

$$FO^{elastic} = \alpha(G(n) - G(n-1))^2 dn \quad (1.5)$$

$$FO^{bend} = \beta(G(n-1) - G(n) + (G(n+1))^2 dn \quad (1.6)$$

Finally, the pre-processed image Im_{pre} is determined from the initial stage.

1.4 Feature Extraction using MPCA and MLDA

1.4.1 MPCA

The pre-processed image Im_{pre} is then subjected to the MPCA [17] model for extracting the features. MPCA is an extended version of PCA and it is also known as 'data tensor'. MPCA includes image reconstruction, by which the image is reorganized into 3D tensor as $\text{Im}_{te} \in K^{\text{Im}_{pre}^1 \times \text{Im}_{pre}^2 \times \text{Im}_{pre}^3}$ in which Im_{pre}^1, Im_{pre}^2 and Im_{pre}^3 refers to the height, width, and count of the images. The explanation of the MPCA model is given below in Figure 1.4:

1.4.2 MLDA

The Linear Discriminant Analysis looks for a straight line later projecting the data on an exact axis to distinct the binary classes [14]. The straight lines unscrambling the twofold classes after the prognosis make the centers of the two classes far away from each other, and make each alteration minor (refer Figure 1.5). y is a one-dimensional vector where a vector x of p-dimension is expected on vector w. For C1 class with N1 data and C2 class with N2 data, the center vector of each class is m1 and m2.

- Calculate the mean matrix $\bar{O} = \left(1/\text{Im}_{pre}^3\right) \sum_{i=1}^{\text{Im}_{pre}^3} \bar{E}_i$;
- The tensor $\hat{\text{Im}}_{te} = \left[\bar{E}_1 - \bar{O}, \bar{E}_2 - \bar{O}, \ldots, \bar{E}_{\text{Im}_{pre}^3} - \bar{O}\right]$ is cantered;

Figure 1.4   Process of a multilinear projection.

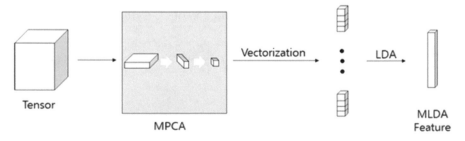

Figure 1.5   Process of a multilinear projection.

- $\hat{\text{Im}}_{te}$ is unfolded into a matrix. The components of the mode *md* unfolding matrix are portrayed as $\left(\hat{\text{Im}}_{te}^{(md)}\right)_{(index)} = b_{i_1 i_2 i_3}$ in which index = $i_{md}, \sum_{o=3}^{md+1} (i_o - 1) + \sum_{o=md-1} (i_o - 1)$;
- By covariance matrix, the eigenvectors are determined by mode-*md* that computes the matrix $K, C^{(m)} = \hat{K}_{(md)} \hat{K}_{(md)}^K \cdot$, and assume the eigenvectors as $B_{(md)} = \left[B_1, B_2, \ldots, B_{LA_{(md)}} \right]$ in which $LA_{(md)}$ refers to the highest eigenvectors;
- The feature selection is of 2 kinds: (a) by the sample $S_{i(md)} = (\bar{E}_i - \bar{O}) \times B_{(md)}^K$ for diverse modes and (b) by $S_i = B_1^P (\bar{E} - \bar{O}) \times B_{(2)}$. There will be various percentages captured for each node;
- The classification is performed on any one of these feature sets.

1.5 Classification

This work exploits NN [18, 24] for recognizing caries. The input feature set is given by Equation (1.7), in which N_D denotes the count of elected features.

$$FE^{weight} = [F_1, F_2, F_3, F_4 \ldots F_{N_D}] \tag{1.7}$$

The weight WE of the network model is portrayed by the LM framework. Equation (1.8) portrays the NN framework, in which the resultant output from i^{th} node of jth layer is given by $ou_l^{(j)}$. The input is signified by $FE^{weight_i^j}$, $af(\bullet)$ indicates the activation function, the entire count of input to j^{th} layer is given by $nu^{(j)}$, bi_i symbolizes the input bias to j^{th} layer, c and d denotes the weight coefficient of WE as specified in Equation (1.9). The predicted network output $\hat{P}$ is given by Equation (1.10), in which w^0 signifies the bias weight and $w^{(h)}$ defines the hidden neuron weight.

$$ou_l^{(j)} = af \left[c_l^{(j)} bi_j + \sum_{i=1}^{nu^{(j)}} FE_i'^{weight(j)} d_{il}^{(j)} \right] \tag{1.8}$$

$$WE = [c; d] \tag{1.9}$$

$$\hat{P} = w^0 + \sum_{i=1}^{nu^{(j)}} ou_l^{(j)} w_i^{(h)} WE \tag{1.10}$$

So as to train the network, the network weight WE^* is optimally chosen with the determination of objective function as in Equation (1.11), where P

indicates the actual output.

$$WE^* = \arg\min [WE] \, ||P - \hat{P}|| \tag{1.11}$$

Thus the classifier classifies the input image (non-caries or caries image).

1.5.1 Classification

For enhancing the adopted scheme, the extracted features are multiplied with the weight W given in Equation (1.12). The weight size should be equal to the attained feature size. The distance D amongst the attained features FE^{weight} is given by Equation (1.13).

$$W = \lfloor W_1, \ldots, W_{N_D} \rfloor \tag{1.12}$$

$$FE^{weight} = FE \times W = FE_1^{weight}, \ldots, FE_{N_D}^{weight} \tag{1.13}$$

The distance di_o, in which $o = 1, 2, 3, \ldots, N_D$ amongst attained f^{weight} is computed by 'Nonlinear programming optimization model'. The objective Obj function is given by Equation (1.14).

$$Obj = \max(di_o) : o = 1, 2, \ldots, N_D \tag{1.14}$$

1.5.2 Nonlinear Programming Optimization

The issue regarding the nonlinear program is given in Equation (1.15), in which $\hat{h}(\hat{x})$, $\hat{i}(\hat{x})$ and $\hat{j}(\hat{x})$ are portrayed as 'deferential functions'.

$$\min_{\hat{y}} \hat{h}(\hat{x}) = 0 \tag{1.15}$$

So that

$$\begin{aligned} \hat{i}(\hat{x}) &= 0 \\ \hat{j}(\hat{x}) &= 0 \end{aligned} \tag{1.16}$$

The substitution of Equation (1.15) is done by a sequence of barrier sub issues as specified in Equation (1.17), in which $\hat{l} > 0$ points out the vector of slack parameters, $\hat{k} = \left(\hat{x}, \hat{l} \right)$ and $\mu > 0$ denotes the barrier constraint.

$$\min_{\hat{k}} \varphi_\mu \left(\hat{k} \right) \equiv \hat{h}(\hat{x}) - \mu \sum_{\hat{o}}^{\hat{n}} In\hat{l}_{\hat{o}} \tag{1.17}$$

$$\hat{i}\left(\hat{y}\right) = 0$$

So that $\quad \hat{j}\left(\hat{x}\right) + \hat{l} = 0 \qquad\qquad (1.18)$

The Lagrangian function associated with Equation (1.17) is specified in Equation (1.19), in which $\zeta_{\hat{i}}, \zeta_{\hat{a}}$ indicates the 'Lagrange multipliers' and $\zeta = \left(\zeta_{\hat{i}}, \zeta_{\hat{a}}\right)$.

$$\aleph\left(\hat{k}, \zeta; \mu\right) = \varphi_\mu\left(\hat{k}\right) + \zeta_{\hat{i}}^{\hat{v}} \hat{i}\left(\hat{x}\right) + \zeta_{\hat{a}}^{\hat{v}}\left(\hat{a}\left(\hat{x}\right) + \hat{l}\right) \qquad (1.19)$$

The optimality states in Equation (1.17) could be specified as per Equation (1.20), in which $\hat{l}$ and $\zeta_{\hat{a}}$ are non-negative, $\hat{Y}_{\hat{i}}$ and $\hat{Y}_{\hat{a}}$ refers to Jacobian matrices, $\hat{D}$ and $\Gamma_{\hat{a}}$ points out the diagonal matrices.

$$\begin{bmatrix} \nabla \hat{h}\left(\hat{x}\right) + \hat{Y}_{\hat{i}}\left(\hat{x}\right)^{\hat{v}} \zeta_{\hat{i}} + \hat{Y}_{\hat{a}}\left(\hat{x}\right)^{\hat{v}} \zeta_{\hat{a}} \\ \hat{D}\Gamma_{\hat{a}}\hat{e} - \mu\hat{e} \end{bmatrix} = \begin{bmatrix} 0 \\ 0 \end{bmatrix} \qquad (1.20)$$

Further, the current iterate $\left(\hat{k}, \zeta\right)$ outcomes in the primal-dual system as given by Equation (1.21), in which $\hat{z}_{\hat{k}} = \begin{bmatrix} \hat{z}_{\hat{x}} \\ \hat{z}_{\hat{l}} \end{bmatrix}$, $\hat{z}_\zeta = \begin{bmatrix} \hat{z}_{\hat{i}} \\ \hat{z}_{\hat{a}} \end{bmatrix}$,

$\hat{c}\left(\hat{k}\right) = \begin{bmatrix} \hat{i}\left(\hat{x}\right) \\ \hat{j}\left(\hat{x}\right) + \hat{l} \end{bmatrix}$, $\hat{Y}\left(\hat{x}\right) = \begin{bmatrix} \hat{Y}_{\hat{i}}\left(\hat{x}\right) & o \\ \hat{Y}_{\hat{a}}\left(\hat{x}\right) & 1 \end{bmatrix}$ and $\hat{R}\left(\hat{k}, \zeta; \mu\right) =$

$\begin{bmatrix} \nabla_{\hat{x}\hat{x}}^2 \aleph\left(\hat{k}, \zeta; \mu\right) & 0 \\ 0 & \hat{D}^{-1}\Gamma_{\hat{a}} \end{bmatrix}$

$$\begin{bmatrix} \hat{R}\left(\hat{k}, \zeta; \mu\right) & \hat{Y}\left(\hat{x}\right)^{\hat{v}} \\ \hat{Y}\left(\hat{x}\right) & 0 \end{bmatrix} \begin{bmatrix} \hat{z}_{\hat{k}} \\ \hat{z}_\zeta \end{bmatrix} = -\begin{bmatrix} \nabla_{\hat{k}} \aleph\left(\hat{k}, \zeta; \mu\right) \\ \hat{c}\left(\hat{k}\right) \end{bmatrix} \qquad (1.21)$$

The novel iteration is specified as given in Equation (1.22), in which $\alpha_{\hat{k}}$ and α_ζ denotes the step-lengths that are modeled as per Equation (1.23).

$$\hat{k}^+ = \hat{k} + \alpha_{\hat{k}} \hat{z}_{\hat{k}}, \quad \zeta^+ = \zeta + \alpha_\zeta \hat{z}_\zeta \qquad (1.22)$$

$$\begin{aligned} \alpha_{\hat{k}}^{\max} &= \max\left[\alpha \in (0,1)\right] : \hat{l} + \alpha\hat{z}_{\hat{l}} \geq (1-\tau)\,\hat{l} \\ \alpha_\zeta^{\max} &= \max\left[\alpha \in (0,1)\right] : \zeta_{\hat{a}} + \alpha\hat{z}_{\hat{a}} \geq (1-\tau)\,\zeta_{\hat{a}} \end{aligned} \qquad (1.23)$$

1.6 Proposed Artificial Dragonfly Algorithm for solving Optimization Problem

In this work, modified ADA is implemented for training the NN classifier. The DA model [21, 23] concerns on five factors for updating the location of the dragonfly. They are (i) Control cohesion (ii) Alignment (iii) Separation (iv) Attraction (iv) Distraction. The separation of rth dragonfly, M_r is calculated by Equation (1.24) and here A denotes the current dragonfly position, A'_s refers to the location of sth neighbouring dragonfly and H' denotes the count of neighboring dragonflies.

$$M_r = \sum_{s=1}^{H'} \left(A' - A'_s \right) \tag{1.24}$$

The alignment and cohesion are computed by Equation (1.25) and Equation (1.26). In Equation (1.25), Q'_s refers to the velocity of sth neighbour dragonfly.

$$J_r = \frac{\sum\limits_{s=1}^{H'} Q'_s}{H'} \tag{1.25}$$

$$V_r = \frac{\sum\limits_{s=1}^{H'} A'_s}{H'} - A \tag{1.26}$$

Attraction towards food and distraction to the enemy are illustrated in Equation (1.27) and Equation (1.28). In Equation (1.27), Fo refers to the food position and in Equation (1.28), ene denotes the enemy position.

$$W_r = Fo - A' \tag{1.27}$$

$$Z_r = ene + A' \tag{1.28}$$

The vectors such as position A' and $\Delta A'$ step are considered here for updating the position of the dragonfly. The step vector $\Delta A'$ denotes the moving direction of dragonflies as given in Equation (1.29), in which q', t', v', u', z' and δ refers the weights for separation, alignment, cohesion, food factor, enemy factor, and inertia respectively and l denotes to the iteration count.

$$\Delta A'_{l+1} = \left(q'M_r + t'J_r + v'V_r + u'W_r + z'Z_r \right) + \delta.\Delta A'_l \tag{1.29}$$

The position vector is computed by Equation (1.30), in which l denotes the present iteration, L^{best} and G^{best} denotes the local and global best solutions respectively and τ indicates the variation of fitness as given by Equation (1.31).

$$A'_{l+1} = A'_l + \Delta A'_{l+1} + \left(L^{best} + G^{best} \right) \times \tau \qquad (1.30)$$

$$\tau = \frac{A(l-1) - A(l)}{A(l)} \qquad (1.31)$$

If there are no neighborhood solutions, the position is updated as per Equation (1.32) and (1.33) here d refers to the size of position vectors, y_1 and y_2 refers to the two arbitrary integers in [0,1], $\wp$ signifies a stable parameter. With the increase in iteration, the position and steps of all dragonflies are updated as per Equations (1.32), (1.33), and (1.34).

$$A'_{l+1} = A'_l + levy\,(d) \times A'_l \qquad (1.32)$$

$$levy\,(d) = 0.01 \times \frac{y_1 \times \Phi}{|y_2|^{\frac{1}{d}}} \qquad (1.33)$$

$$\Phi = \left(\frac{\Psi\,(1 + \wp) \times \sin\left(\frac{\pi \wp}{2}\right)}{\psi\left(\frac{1+\wp}{2}\right) \times \wp \times 2^{\left(\frac{\wp-1}{2}\right)}} \right) \qquad (1.34)$$

$$\psi\,(m) = (m-1)\,!$$

At last, the algorithm provides the optimized weight that is multiplied with the features that are extracted, which is then classified using NN for detecting caries.

1.7 Results and Discussion

The analysis on dental carried recognition was simulated using MATLAB. The dataset was attained from https://mynotebook.labarchives.com that was defined into 2 test cases, which were necessary to offer better outcomes. Every set includes 60 caries images. Accordingly, the algorithmic analysis was carried out with respect to 'accuracy, sensitivity, specificity, FPR, FNR, FDR, NPV, FOR, BM, MK, MCC, and F1Score' by varying the percentage of variation (T) captured for each node in MPCA. For analysis purpose, T was varied from 94, 95, 97, 98, 99 and 100 with respect to cohesion weight v' that was varied between 0.1, 0.2, 0.4, 0.6, 0.7 and 0.9.

1.8 Result Interpretation

The presentation scrutiny of the implemented model with respect to varied values of T is given by Figures 1.6–1.8 and 1.9 for accuracy, sensitivity, specificity, and F1 Score respectively. For instance, from Figure 1.6 accuracy of T at 97 is high, which is 3.06%, 3.06%, 8.16%, and 6.12% better than T at 94, 95, 98, 99, and 100 when v' is 0.2. From Figure 1.6, the accuracy of the adopted model when $T = 95$ is high, which is 8.16%, 13.27%, 8.16% and 16.33% better than T at 97, 98, 99 and 100 when v' is 0.4. On considering Figure 1.6, the accuracy at $T = 95$ is high, which is 7.53%, 3.23%, 3.23% and 3.23% better than T at 97, 98, 99 and 100 when v' is 0.2. Likewise, from Figure 1.7, the sensitivity of the adopted scheme when $T = 97$ is higher, which is 1.08%, 2.15%, 1.08%, and 16.13% better than T at 94, 95, 98,

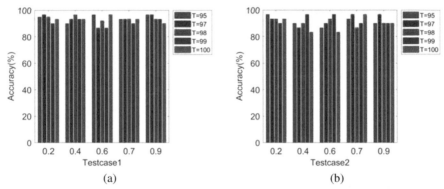

Figure 1.6   Accuracy analysis of the adopted model by varying (a) test case 1 (b) test case 2.

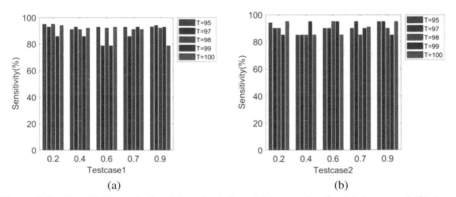

Figure 1.7   Sensitivity analysis of the adopted model by varying for (a) test case 1 (b) test case 2.

99 and 100 when v' is 0.9. Also, from Figure 1.7, the sensitivity at $T = 97$ is more, which is 7.22%, 12.37%, 7.22% and 6.19% better than T at 95, 98, 99 and 100 when v' is 0.7. Moreover, Figure 1.8 shows the specificity of the adopted model, which revealed better results for all the two test cases. From Figure 1.8, the specificity of the presented model at $T = 95$ is high, which is 3.23%, 8.6%, 8.6%, and 8.6% better than T at 97, 98, 99 and 100 when v' is 0.7. From Figure 1.8, the specificity of the presented model at $T = 99$ is high, which is 13.04%, 2.17%, 2.17% and 13.04% better than T at 95, 97, 98 and 100 when v' is 0.6. From Figure 1.8, the specificity when $T = 99$ is high, which is 21.05%, 21.05%, 47.37% and 47.37% better than T at 95, 97, 98 and 100 when v' is 0.7. The F1-score of the adopted model is revealed by

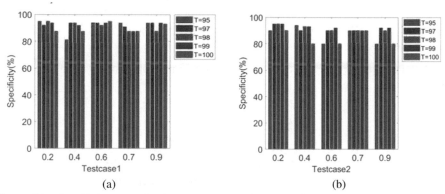

Figure 1.8 Specificity analysis of the adopted model by varying for (a) test case 1 (b) test case 2.

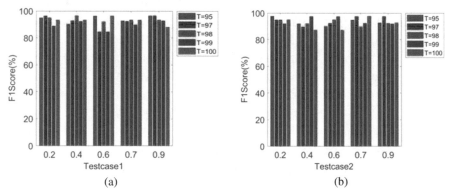

Figure 1.9 F1-Score analysis of the adopted model by varying for (a) test case 1 (b) test case 2.

Figure 1.9, which shows betterment for all values of T. From Figure 1.9, the F1-score of the implemented model at $T = 95$ is high, which is 3.23%, 8.6%, 8.6% and 8.6% better than T at 97, 98, 99 and 100 when v' is 0.4. From Figure 1.9, the F1-score at $T = 99$ is high, which is 3.23%, 8.6%, 8.6% and 8.6% better than T at 95, 97, 98 and 100 when v' is 0.4. Thus, the betterment of the adopted scheme has been validated effectively.

1.9 Performance Analysis by Varying Learning Percentage

The performance analysis under different variations of learning percentages for 2 test cases is given in Table 1.1 through Table 1.6 for varied values. Here, the learning percentage is varied from 15%, 30%, 45%, and 95% for the three test cases.

Table 1.1   Performance analysis with respect to learning percentage of 15% and 30% for database 1

Training Percentage	15%					30%				
Varying Values of T	95	97	98	99	100	95	97	98	99	100
Sensitivity	0.89	0.89	0.94	0.94	0.94	0.95	0.91	0.93	0.93	0.93
Accuracy	0.94	0.92	0.97	0.92	0.94	0.95	0.90	0.97	0.93	0.97
Precision	0.95	0.94	0.94	0.89	0.94	0.95	0.82	0.94	0.93	0.93
Specificity	0.95	0.94	0.94	0.89	0.94	0.95	0.81	0.94	0.94	0.94
FNR	0.11	0.11	0.06	0.06	0.06	0.13	0.13	0.07	0.07	0.13
FPR	0.13	0.06	0.13	0.11	0.06	0.13	0.19	0.13	0.06	0.06
FDR	0.13	0.06	0.13	0.11	0.06	0.13	0.18	0.13	0.07	0.07
NPV	0.95	0.94	0.94	0.89	0.94	0.95	0.81	0.94	0.94	0.94
F1score	0.94	0.91	0.97	0.92	0.94	0.95	0.90	0.96	0.93	0.97
MCC	0.89	0.83	0.95	0.83	0.89	0.95	0.82	0.93	0.87	0.94
FOR	0.13	0.06	0.13	0.11	0.06	0.13	0.19	0.13	0.06	0.06
BM	0.89	0.83	0.94	0.83	0.89	0.95	0.81	0.93	0.87	0.94
MK	0.95	0.89	0.94	0.78	0.89	0.95	0.64	0.94	0.87	0.87

Table 1.2 Performance analysis with respect to learning percentage of 60% and 80% for database 1

Training Percentage	60%					80%				
Varying Values of T	95	97	98	99	100	95	97	98	99	100
Sensitivity	0.90	0.90	0.91	0.95	0.90	0.80	0.90	0.94	0.60	0.90
Accuracy	0.90	0.95	0.91	0.85	0.90	0.80	0.90	0.90	0.70	0.90
Precision	0.90	0.94	0.91	0.77	0.83	0.80	0.90	0.83	0.75	0.83
Specificity	0.90	0.94	0.91	0.70	0.80	0.80	0.90	0.80	0.80	0.80
FNR	0.10	0.10	0.13	0.13	0.13	0.20	0.13	0.13	0.40	0.13
FPR	0.10	0.13	0.13	0.30	0.20	0.20	0.13	0.20	0.20	0.20
FDR	0.10	0.13	0.13	0.23	0.17	0.20	0.13	0.17	0.25	0.17
NPV	0.90	0.94	0.91	0.70	0.80	0.80	0.90	0.80	0.80	0.80
F1score	0.90	0.95	0.91	0.87	0.91	0.80	0.90	0.91	0.67	0.91
MCC	0.80	0.90	0.91	0.73	0.82	0.60	0.90	0.82	0.41	0.82
FOR	0.10	0.13	0.13	0.30	0.20	0.20	0.13	0.20	0.20	0.20
BM	0.80	0.90	0.91	0.70	0.80	0.60	0.90	0.80	0.40	0.80
MK	0.80	0.94	0.91	0.47	0.63	0.60	0.90	0.63	0.55	0.63

Table 1.3 Performance analysis with respect to learning percentage of 10% and 25% for database 2

Training Percentage	10%					25%				
Varying Values of T	95	97	98	99	100	95	97	98	99	100
Sensitivity	0.87	0.87	0.87	0.87	0.87	0.94	0.85	0.90	0.90	0.95
Accuracy	0.92	0.92	0.92	0.92	0.92	0.97	0.90	0.87	0.93	0.90
Precision	0.94	0.91	0.91	0.93	0.94	0.90	0.94	0.80	0.90	0.80
Specificity	0.13	0.13	0.13	0.13	0.13	0.10	0.13	0.20	0.13	0.20
FNR	0.94	0.91	0.91	0.93	0.94	0.95	0.94	0.90	0.90	0.90
FPR	0.94	0.91	0.91	0.93	0.94	0.90	0.94	0.80	0.90	0.80
FDR	0.13	0.13	0.13	0.13	0.13	0.13	0.15	0.10	0.10	0.05
NPV	0.13	0.13	0.13	0.13	0.13	0.05	0.13	0.10	0.13	0.10
F1score	0.93	0.93	0.93	0.93	0.93	0.98	0.92	0.90	0.95	0.93
MCC	0.84	0.84	0.84	0.84	0.84	0.93	0.81	0.70	0.87	0.77
FOR	0.13	0.13	0.13	0.13	0.13	0.10	0.13	0.20	0.13	0.20
BM	0.87	0.87	0.87	0.87	0.87	0.90	0.85	0.70	0.90	0.75
MK	0.94	0.91	0.91	0.93	0.94	0.85	0.94	0.70	0.90	0.70

Table 1.4   Performance analysis with respect to learning percentage of 60% and 80% for database 2

Training Percentage	60%					80%				
Varying Values of T	95	97	98	99	100	95	97	98	99	100
Sensitivity	0.92	0.92	0.75	0.94	0.95	0.89	0.89	0.90	0.89	0.94
Accuracy	0.95	0.92	0.80	0.90	0.95	0.90	0.90	0.90	0.90	0.94
Precision	0.94	0.92	0.92	0.94	0.95	0.90	0.95	0.90	0.92	0.94
Specificity	0.75	0.92	0.92	0.75	0.95	0.90	0.95	0.13	0.92	0.94
FNR	0.13	0.13	0.25	0.06	0.13	0.11	0.11	0.13	0.11	0.13
FPR	0.25	0.13	0.13	0.25	0.13	0.13	0.13	0.90	0.13	0.13
FDR	0.75	0.92	0.92	0.75	0.95	0.90	0.95	0.13	0.92	0.94
NPV	0.06	0.13	0.13	0.06	0.13	0.13	0.13	0.10	0.13	0.13
F1score	0.97	0.92	0.86	0.94	0.95	0.94	0.94	0.95	0.94	0.94
MCC	0.84	0.92	0.61	0.69	0.95	0.67	0.67	0.46	0.67	0.94
FOR	0.25	0.13	0.13	0.25	0.13	0.13	0.13	0.90	0.13	0.13
BM	0.75	0.92	0.75	0.69	0.95	0.89	0.89	0.13	0.89	0.94
MK	0.69	0.92	0.92	0.69	0.95	0.90	0.95	0.10	0.92	0.94

Table 1.5   Performance analysis with respect to learning percentage of 20% and 35% for database 2

Training Percentage	20%					35%				
Varying Values of T	95	97	98	99	100	95	97	98	99	100
Sensitivity	0.93	0.91	0.93	0.93	0.96	0.96	0.92	0.88	0.94	0.88
Accuracy	0.89	0.94	0.94	0.86	0.89	0.93	0.93	0.90	0.97	0.83
Precision	0.75	0.75	0.91	0.63	0.63	0.75	0.92	0.92	0.75	0.50
Specificity	0.93	0.93	0.91	0.90	0.90	0.96	0.92	0.92	0.96	0.92
FNR	0.07	0.13	0.07	0.07	0.04	0.04	0.08	0.12	0.13	0.12
FPR	0.25	0.25	0.13	0.38	0.38	0.25	0.13	0.13	0.25	0.50
FDR	0.07	0.07	0.13	0.10	0.10	0.04	0.13	0.13	0.04	0.08
NPV	0.75	0.75	0.91	0.63	0.63	0.75	0.92	0.92	0.75	0.50
F1score	0.93	0.97	0.96	0.91	0.93	0.96	0.96	0.94	0.98	0.90
MCC	0.68	0.84	0.86	0.58	0.66	0.71	0.78	0.71	0.85	0.35
FOR	0.25	0.25	0.13	0.38	0.38	0.25	0.13	0.13	0.25	0.50
BM	0.68	0.75	0.93	0.55	0.59	0.71	0.92	0.88	0.75	0.38
MK	0.68	0.68	0.91	0.52	0.53	0.71	0.92	0.92	0.71	0.42

Table 1.6 Performance analysis with respect to learning percentage of 45% and 70% for database 2

Training Percentage	45%					70%				
Varying Values of T	95	97	98	99	100	95	97	98	99	100
Sensitivity	0.94	0.94	0.89	0.90	0.91	0.90	0.91	0.93	0.93	0.90
Accuracy	0.95	0.90	0.90	0.95	0.91	0.90	0.91	0.93	0.93	0.90
Precision	0.91	0.50	0.95	0.50	0.91	0.46	0.46	0.46	0.46	0.46
Specificity	0.13	0.50	0.13	0.50	0.13	0.46	0.46	0.46	0.46	0.46
FNR	0.91	0.94	0.95	0.95	0.91	0.91	0.91	0.93	0.93	0.91
FPR	0.91	0.50	0.95	0.50	0.91	0.46	0.46	0.46	0.46	0.46
FDR	0.06	0.06	0.11	0.13	0.13	0.10	0.13	0.13	0.13	0.10
NPV	0.13	0.06	0.13	0.05	0.13	0.13	0.13	0.13	0.13	0.13
F1score	0.97	0.94	0.94	0.97	0.91	0.95	0.91	0.93	0.93	0.95
MCC	0.79	0.44	0.67	0.69	0.91	0.46	0.46	0.46	0.46	0.46
FOR	0.13	0.50	0.13	0.50	0.13	0.46	0.46	0.46	0.46	0.46
BM	0.94	0.44	0.89	0.50	0.91	0.46	0.46	0.46	0.46	0.46
MK	0.91	0.44	0.95	0.45	0.91	0.46	0.46	0.46	0.46	0.46

1.10 Conclusion

This chapter has intended a unique caries recognition outline that comprises three segments. At the preliminary phase, the excellence of the image was amended by pre-processing arrangements like CLAHE, contrast enhancement, grey thresholding, and active contour. Furthermore, the topographies get mined using MPCA as well as MLDA and subsequently cataloging was conceded out using augmented NN, where Artificial DA prototypical was projected for the training procedure. Lastly, an Interpretative investigation was accomplished to authenticate the enactment of the projected model. From the investigation, the accuracy of the implemented model when $T = 95$ was high, which was 8.16%, 13.27 8.16%, and 16.33% better than at 97, 98, 99 and 100 when is 0.4. Moreover, the accuracy at $T = 95$ was high, which was 7.53%, 3.23%, 3.23%, and 3.23% better than at 97, 98, 99, and 100 when was 0.2. Thus, the enhancement of the implemented system has been confirmed commendably from the accomplished outcomes.

References

[1] A. Dündar, M. Ertuğrul Çiftçi, Ö. İşman, A. M. Aktan, 'In vivo performance of near-infrared light trans illumination for dentine proximal caries detection in permanent teeth', The Saudi Dental Journal, In press, corrected proof, Available online 28 August 2019.

[2] D. D. Divakar, S. Ali S. Alanazi, Md Yahya A. Assiri, S. Md Halawani, Md Mustafa, 'Association between ENAM polymorphisms and dental caries in children', Saudi Journal of Biological Sciences, vol. 26, no. 4, pp. 730–735, May 2019.

[3] H. Kang, J. J. Jiao, C. Lee, M. H. Le, C. L. Darling and D. Fried, 'Nondestructive Assessment of Early Tooth Demineralization Using Cross-Polarization Optical Coherence Tomography', IEEE Journal of Selected Topics in Quantum Electronics, vol. 16, no. 4, pp. 870–876, Aug. 2010.

[4] A. Sampathkumar, D. A. Hughes, K. J. Kirk, W. Otten and C. Longbottom, 'All-optical photoacoustic imaging and detection of early-stage dental caries', IEEE International Ultrasonic Symposium, 2014.

[5] P. Joris et al., 'Pre-processing of Heteroscedastic Medical Images,' IEEE Access, vol. 6, pp. 26047–26058, 2018.

[6] J. Sultana, M. S. Islam, M. R. Islam and D. Abbott, 'High numerical aperture, highly birefringent novel photonic crystal fibre for medical imaging applications', Electronics Letters, vol. 54, no. 2, pp. 61–62, 2018.

[7] K. Angelino, D. A. Edlund and P. Shah, 'Near-Infrared Imaging for Detecting Caries and Structural Deformities in Teeth', IEEE Journal of Translational Engineering in Health and Medicine, vol. 5, pp. 1–7, 2017.

[8] R. C. Lee, M. Staninec, O. Le and D. Fried, 'Infrared Methods for Assessment of the Activity of Natural Enamel Caries Lesions', IEEE Journal of Selected Topics in Quantum Electronics, vol. 22, no. 3, pp. 102–110, May-June 2016.

[9] H. Kang, J. J. Jiao, C. Lee, M. H. Le, C. L. Darling and D. Fried, 'Nondestructive Assessment of Early Tooth Demineralization Using Cross-Polarization Optical Coherence Tomography', IEEE Journal of Selected Topics in Quantum Electronics, vol. 16, no. 4, pp. 870–876, 2010.

[10] M. Lashgari, M. Shahmoradi, H. Rabbani and M. Swain, 'Missing Surface Estimation Based on Modified Tikhonov Regularization: Application for Destructed Dental Tissue', IEEE Transactions on Image Processing, vol. 27, no. 5, pp. 2433–2446, May 2018.

[11] B. Winstone, C. Melhuish, T. Pipe, M. Callaway and S. Dogramadzi, 'Toward Bio-Inspired Tactile Sensing Capsule Endoscopy for Detection of Submucosal Tumors', IEEE Sensors Journal, vol. 17, no. 3, pp. 848–857, Feb. 2017.

[12] C. B. Top, A. K. Tafreshi and N. G. GenÃğer, 'Microwave Sensing of Acoustically Induced Local Harmonic Motion: Experimental and Simulation Studies on Breast Tumor Detection', IEEE Transactions on Microwave Theory and Techniques, vol. 64, no. 11, pp. 3974–3986, 2016.

[13] R. Kaur and S. Kaur, 'Comparison of contrast enhancement techniques for medical image', 2016 Conference on Emerging Devices and Smart Systems (ICEDSS), Namakkal, pp. 155–159, 2016.

[14] H. Lu, Konstantinos N. (Kostas) Plataniotis and A. N. Venetsanopoulos, 'MPCA: Multilinear Principal Component Analysis of Tensor Objects', IEEE Transactions on Neural Networks, vol. 19, no 1, 2008.

[15] Y. Mohan, S. S. Chee, D. K. Pei Xin and L. P. Foong, 'Artificial Neural Network for Classification of Depressive and Normal in EEG', IEEE EMBS Conference on Biomedical Engineering and Sciences, 2016.

[16] M. Bakoš,' 5th Slovakian – Hungarian Joint Symposium on Applied Machine Intelligence and Informatics', January 2007.

[17] Hetal J. Vala and Astha Baxi,' A Review on Otsu Image Segmentation Algorithm', International Journal of Advanced Research in Computer Engineering & Technology, vol. 2, no. 2, 2013.

[18] Md Jafari, Md Hossein B. Chaleshtari, 'Using dragonfly algorithm for optimization of orthotropic infinite plates with a quasi-triangular cut-out', European Journal of Mechanics - A/Solids, vol. 6, pp. 1–146, Dec. 2017.

[19] S. Nirkhi, S. Patil, 'Comprehensive Assessment of Imbalanced Data Classification', International Journal of Engineering and Advanced Technology, vol. 9, no. 4, pp. 1426–1431, 2020. doi: 10.35940/ijeat .d7349.049420

[20] S. Patil, S. N., 'Deep Learning Techniques for Oral Diagnosis and Cavity Recognition: A systematic Approach', International Journal of Advanced Science and Technology, vol. 29, no. 9, pp. 192–199, 2020.

[21] S. Patil, V. Kulkarni, and A. Bhise, 'Algorithmic analysis for dental caries detection using an adaptive neural network architecture', Heliyon, vol. 5, no. 5, 2019. doi:10.1016/j.heliyon.2019.e01579

[22] S. Patil, V. Kulkarni, and A. Bhise, 'BEASF-Based Image Enhancement for Caries Detection Using Multidimensional Projection and Neural

Network', International Journal of Artificial Life Research, vol. 8, no. 2, pp. 47–66, 2019. doi:10.4018/ijalr.2018070103

[23] S. Patil, V. Kulkarni, and A. Bhise, 'Caries Detection with the Aid of Multilinear Principal Component Analysis and Neural Network', In Proceedings of the 2nd International Conference on Green Computing and Internet of Things, ICGCIoT, Institute of Electrical and Electronics Engineers Inc., pp. 272–277, 2018. doi: 10.1109/ICGCIo T.2018.8753002

[24] S. Patil, V. Kulkarni, and A. Bhise, 'Caries detection using multidimensional projection and neural network', International Journal of Knowledge-Based and Intelligent Engineering Systems, vol. 22, no. 3, pp. 155–166, 2018. doi:10.3233/KES-180381

[25] S. Patil, V. Kulkarni, and A. Bhise, 'Comprehensive Assessment of Dental Cone Beam Computed Tomography (CBCT): A Systematic Approach', Test Engineering & Management, vol. 83, pp. 16243–16256, May-June 2020. ISSN: 0193–4120.

[26] Jae-Hong Lee, Do-Hyung Kim, Seong-Nyum Jeong, Seong-Ho Choi, 'Detection and diagnosis of dental caries using a deep learning-based convolutional neural network algorithm', Journal of Dentistry, vol. 77, pp. 106–111, October 2018.

[27] Y. Wu, E. C. Jansen, K. E. Peterson, B. Foxman, E. A. Martinez-Mier, 'The associations between lead exposure at multiple sensitive life periods and dental caries risks in permanent teeth', Science of The Total Environment, vol. 654, pp. 1048–1055, March 2019.

[28] C. Signori, M. Laske, E. M. Bronkhorst, Marie-Charlotte D. N. J. M. Huysmans, N. J. M. Opdam, 'Impact of individual-risk factors on caries treatment performed by general dental practitioners', Journal of Dentistry, vol. 81, pp. 85–90, February 2019.

2

Lung Cancer Prediction using Feature Selection and Recurrent Residual Convolutional Neural Network (RRCNN)

Syed Saba Raoof[1], M. A. Jabbar[2], and Syed Aley Fathima[2]

[1]VIT University, India
[2]Vardhaman College of Engineering, India
E-mail: syedsabaraoof@gmail.com; jabbar.meerja@gmail.com;
noorainmustafa786@gmail.com

Abstract

Lung cancer (LC) is the most prevailing cause of cancer deaths globally. Detecting LC by using Computed Tomography (CT) images is the predominant method. Over recent years, Deep learning (DL) techniques have a profound impact on providing the best performance in various fields of research. These techniques are successfully implemented in the medical imaging field. The U-Net model is the most used and practiced algorithm for medical imaging applications. In the present study, we proposed the recurrent Residual Convolutional Neural Network (RRCNN) model on basis of UNet models, which automatically segment and classify the lung nodules. The proposed model makes use of three underlying algorithms i.e., UNet, Residual Network, as well as Recurrent Convolutional Neural Network (RCNN). For experimental analysis, the LUNA16 dataset had been utilized and the outcome of the model demonstrates that it can efficiently accomplish tasks like detection, identification, segmentation, and classification from the input given compared to identical models along with UNet and RCNN and achieved an accuracy of around 97%.

Key Words: Lung cancer (LC), CT images, LUNA16, UNet, RCNN, and RRCNN.

2.1 Introduction

Cancer is the ruling mainspring of the deaths of humans worldwide. Among all the cancers LC is the largest prevailing and death-dealing cancer. The elite way to rescue from the deaths of LC is earlier diagnosing and felicitous treatment. According to UK Cancer Research, the LC rate of survivability among all the cancers is second-lowest for one year post-prognosis is 30% survival rate and for ten years post-prognosis is only 5% [1–3]. A lung cancer diagnosis can be done through various methods like Computer Tomography (CT), X-ray, Positron Emission Tomography (PET), Magnetic Resonance Imaging (MRI), ultrasound, and isotope, among all the tests X-ray and CT images, are the most well-known and used methods for detection and identification of lung diseases [27].

Deep Learning techniques can be implemented for medical imaging classification, detection, segmentation, and diagnosis. A handful of deep learning algorithms have been developed alike, ResidualNet, AlexNet, GoogleNet, DenseNet, VGG, CapsuleNet, and so on., [42, 43] Convolutional Neural Network (CNN) a DL method improves and eases the task of detection, segmentation and classification by employing various activation and dropout functions which clinch the problems raised during, training the model and regularize the network models. Computer-Aided-Diagnosis (CAD) aims to achieve an agile and enhanced accurate diagnosis model for higher-quality treatment of people at once [33, 34]. Present research work describes various DL algorithms implemented for predicting LC.

Before DL innovation, various traditional ML and Digital Image Processing (DIP) approaches are existing solutions for segmentation, classification, and detection of medical images like Support Vector Machine (SVM), Random Forest (RF), K-nearest neighbor (Knn), decision trees, bayesian classifier, etc.

However, DL is not only restricted to medical imagining, the different DL based techniques are implemented in distinct fields including data acquisition, segmentation, classification, automatic labeling and captioning, computer-aided detection and diagnosis, reading assistants and automatic dictation, advanced Electronic Health Recording (EHR) and precision imaging for personalized medicine. [32] Feature selection has been a widely used dimensionality reduction method used to remove unnecessary features and to increase accuracy.

In this research study, we are concerned with medical image segmentation, detection, and classification. Proposed model had been implemented based on DL techniques for Lung cancer CT images (medical image) segmentation, identification, and classification. An enhanced segmentation model had been implemented by employing Residual-CNN for segmenting the lung nodules. Segmentation is performed by RU-Net and R2U-Net models. The outcome of the proposed system compared by existing models illustrates the best performance. A major contribution of our research study is outlined below:

1 Proposed algorithm is implemented for classifying lung nodules.
2 Proposed network is UNet based including residual connectivity between recurrent convolutional layers named as RRCNN.
3 Concatenation is employed for feature maps from encoding unit to decoding unit.
4 When compared to existing models proposed model performs best and accurately.
5 We applied feature selection to remove redundant features.

The paper is summarized as follows: Section 2.2 describes the literature review. Proposed methodology and system architecture are present in Section 2.3, dataset description and performance measures are described in Section 2.4, and conclusion and future work are illustrated in Section 2.5.

2.2 Related Work

A comprehensive review of various DL approaches has been done and existing methods for detecting and diagnosing cancer is discussed.

Siddhartha Bhatia et al. [4], implemented a model to predict the lung lesion from CT scans by using Deep Convolutional Residual techniques. Various classifiers like XGBoost and Random Forest are used to train the model. Preprocessing is done and feature extraction is done by implementing UNet and ResNet models. LIDC-IRDI dataset is utilized for evaluation and 84% of accuracy is recorded.

A. Asuntha et al. [5], implemented an approach to detect and label the pulmonary nodules. Novel deep learning methods are utilized for the detection of lung nodules. Various feature extraction techniques are used then feature selection is done by applying the Fuzzy Particle Swarm Optimization (FPSO) algorithm. Finally, classification is done by Deep learning methods. FPSOCNN is used to reduce the computational problem of CNN. Further valuation is done on a real-time dataset collected from Arthi Scan Hospital. The experimental analysis determines that the novel FPSOCNN gives the best results compared to other techniques.

Fangzhou Lia et al. [6], developed a 3D deep neural network model which comprises of two modules one is to detect the nodules namely the 3D region proposal network and the other module is to evaluate the cancer probabilities, both the modules use a modified U-net network. 2017 Data Science Bowl competition the proposed model won first prize. The overall model achieved better results in the standard competition of lung cancer classification.

Qing Zeng et al. [7]. implemented three variants of DL algorithms namely, CNN, DNN, and SAE. The proposed models are applied to the Ct scans for the classification and the model is experimented on the LIDC-IDRI dataset and achieved the best performance with 84.32% specificity, 83.96% sensitivity and accuracy is 84.15%.

Md Z Alom et al. [8], the author developed a Recurrent CNN (RCNN) UNet based model namely RU-Net model, and a Recurrent Residual CNN (R2CNN) UNET based model namely, R2U-Net. LUNA16 Dataset is utilized for training the model. The outcome of the model demonstrates that the proposed models achieve better performance in segmentation tasks with equal network parameters, compared to existing models i.e., the UNet and ResNet models.

Petros-Pavlos et al. [9] developed deep artificial neural network architecture, ReCTnet for pulmonary nodule detection from Ct scans. The dataset used for experimental analysis is LIDC/IDRI which consists of 1,018 annotated CT scans. The proposed model achieves 90.5% sensitivity at 4.5 average of false positive/scan. This model when compared with Multichannel CNN for multislice segmentation and other models using the same dataset determines that it shows high performance.

Lakshmanaprabu et al [10]; developed an Optimal Deep Neural Network (ODNN) and Linear Discriminate Analysis (LDA) based classification model for CT images. Lung nodule classification is done by applying LDR and optimization is done by applying the Modified Gravitational Search Algorithm to predict lung cancer. Standard CT database is used for experimental analysis which comprises 50-low dosage lung cancer CT images. This model is correlated with existing models such as KNN, NN, DNN SVM, and so on, and the experimental analysis shows the best results for the developed model with 94.56% accuracy, sensitivity, and specificity 96.2% and 94.2% respectively.

Worawate et al [11]; developed a method "Automatic Lung Cancer Prediction from Chest X-ray Images using Deep Learning Approach", Authors used DensNet-121 (121 layers Convolutional neural network) in conjunction with transfer learning for classifying using chest images. Model is trained on two datasets i.e. Chest X-ray 14 and JSRT to identify the nodules. The model obtained an accuracy of about 74.43±6.01%, sensitivity, and specificity of about 74.68 ± 15.33% and 74.96 ± 9.85% respectively.

Jason et al [12]; developed a Deep Screener algorithm which is a form of different deep learning approaches. The Deep Screener is an end-to-end automated screening of lung cancer on low dose CT images. TCIA dataset is used for experimental analysis which comprises 1449 low dose CT images. The model developed was correlated with the grt123 algorithm of Data Science Bowl 2017 for lung cancer analysis and the result was too close wining algorithm grt123. The proposed model predicted about 1359 out of 1449 CT scans with about 82% of accuracy, AUC about 0.885, AUPRC about 0.837.

Christoph et al [13]; proposed a prediction model based on tomography lung cancer images. A CNN is utilized for feature extraction by fine-tuning pre-trained ResNet18 and multimodal features CNN is trained by the Cox model for hazard prediction. Lung1 dataset is used for experimental analysis which can be accessed from "The Cancer Imaging Archive" (TCIA) that comprises 422 NSCLC (Non-Small Cell-Lung Carcinoma) images for 318 of 422 patients.

Ahmed et al [14]; proposed a Computer-Aided Detection model (CADe), which is developed to predict the lung nodules at the initial stage from Low Dose Computed Tomography images (LDCT). Various deep learning algorithms like Alex, VGG19, and VGG16 are used for feature extraction. Experimental results are carried on 320 LDCT images. The proposed architecture, achieved 96.25%, 97.5%, and 95% accuracy, sensitivity, and specificity respectively.

Brahim et al [15]; developed UNet architecture to detect lung nodules from CT scans. The (LIDC-IDRI) includes marked annotated lesions. The proposed architecture consists of two paths namely contracting and symmetric expanding path, to acquire higher level data, and to recover the necessary information respectively. Around 0.9502% of the dice-Coefficient index and precise segmentation accuracy is obtained.

Atsushi et al [16]; a Deep CNN was developed for classifying lung nodules automatically. Dataset is collected from exfoliative cytology which consists of 76 images. Convolutional layers-3, fully connected layers-2, and pooling layers-3 are employed for training. Three types of cancer possibilities were detected using the model and a three-fold cross-validation technique was used for evaluating accuracy. Around 71% of accuracy was achieved

Kuruvilla et al [17]; developed an ANN (Artificial Neural Network) Model based on texture and shape features, and recorded 93.3% of accuracy. Using shape and texture feature combinations for classification, segmentation, and detection improves the overall accuracy for all the tasks.

Hao Wang et al [18]; proposed an LDA approach based on the Euclidean method namely the ELDA approach to improve the existing method and to eradicate the disadvantages in the conventional LDA method. For classification purposes, Multi-class SVM is been used. The experimental results had shown that the proposed algorithm achieved enhanced results similar to improved accuracy compared to other recognition procedures in the model.

Hiba Chougrad et al. [19]; developed CAD CNN based model for classifying breast cancer. Usually, a DL method needs large datasets to train the models, compared to transfer learning models that utilize small datasets. The proposed CNNs are ideally trained by using transfer learning techniques. Around 98.94% of accuracy is achieved by the proposed model.

Cascio et al. [20]; developed a CAD model for cancer classification. CT scans were utilized for experimental analysis. The proposed model achieved a detection rate of about 85% with 6.6 FP/scan (i.e. False Positives). Around 2.47 FPs/CT reductions are achieved at 80%.

Fausto et al. [21]; proposed a model for segmenting the 3D image which relies on volumetric, neural network, and fully convolutional networks. The proposed CNN is trained end-to-end on MRI images depicting the prostate, and the proposed method predicts the segmentation for the entire volume promptly. Dice overlap coefficient is used for analyzing the experimental results and achieved an accuracy of around 83% to 84%.

Zifeng Wu et al. [22] developed a ResNet system for visual recognition. A group of shallow networks was proposed first which had performed well on ImageNet classification. A Fully Connected Network (FCN) was initialized for training the models and semantic image segmentation technique was used for tuning the model. Best accuracy is achieved on four datasets i.e., PASCAL, PASCAL-Context, Cityscapes, ADE20k, and VOC.

Olaf et al. [23] proposed the UNet model for biomedical image segmentation. This model is demonstrated on three distinct segmentation tasks i.e., the EM segmentation task which was started at ISBI 2012. Using the same proposed network transmitted light microscopy images was trained. The proposed architecture achieved an accuracy of around 92.03%.

Comparative study of literature review is summarized in the following Table 2.1.

Table 2.1 Summary of Lung Cancer prediction & detection using various methods and algorithms

S.No.	Author	Method	Dataset	Measures
1	Siddhartha Bhatia et al. [4]	UNet and ResNet	LIDC-IDRI	84% accuracy
2	A. Asuntha et al. [5]	FPOSCNN	the real-time dataset collected from Arthi Scan Hospital	NA
3	Fangzhou Lia et al. [6]	3D deep neural network	LUNA16	Achieved better results
4	QingZeng Song et al. [7]	CNN, DNN and SAE	LIDC-IDRI	84.15%.
5	Md Zahangir Alom et al. [8]	RU-NEet and R2U-Net	LUNA16	–
6	Petros-Pavlos et al. [9]	ReCTnet	LIDC-IDRI	90.5% of sensitivity and 4.5 false positive/scan.
7	Lakshmanaprabu et al, [10]	Optimal deep neural network and Linear Discriminate Analysis	Standard CT dataset with 50 CT images.	Accuracy-94.56%, sensitivity-96.2% & specifity-94.2%
8	Worawate et al. [11]	DenseNet-121	Chest X-ray 14 & JSRT dataset	Accuracy-74.43±6.01, sensitivity-74.68±15.33% & specificity-74.96±9.85%.
9	Jason et al. [12]	Deep Screener algorithm	TCIA-1449 low-dose CT images	82% of accuracy
10	Christoph et al. [13]	ResNet18 & CNN	Lung1 dataset-TCIA	CNN is much harder to interpret when compared with the Cox model. 0.623±0.039 prediction
11	Ahmed et al. [14]	CADe	LDCT	The proposed system, achieved 96.25%, 97.5%, and 95% accuracy, sensitivity, and specificity respectively.

(Continued)

Table 2.1   (*Continued*)

S.No.	Author	Method	Dataset	Measures
12	Brahim et al. [15]	u-net	LIDC-IDRI	95.02%
13	Atsushi et al. [16]	DCNN	Dataset is collected from exfoliative cytology which consists of 76 images.	71% of accuracy.
14	Kuruvilla et al. [17]	ANN	LIDC	Accuracy rate 93.30%.
15	Hao Wang et al. [18]	LDA method based on Euclidean method called ELDA	Ct images	-
16	Hiba Chougrad et al. [19]	CAD-based on CNN	Small dataset	98.94%
17	Cascio et al. [20]	CAD system	CT images	80%
18	Fausto et al. [21]	CNN	MRI prostate images	83% to 84%
19	Zifeng Wu et al. [22]	ResNet	PASCAL, PASCAL-Context, Cityscapes, ADE20k, and VOC.	–
20	Olaf et al. [23]	u-net	Microscopic images	92.30%

2.3 Methodology

As per the literature survey done many researchers had implemented distinct applications in the medical imaging field to detect various diseases automatically which is usually known as Computer-Aided Detection (CAD) system. The main function of medical images in radiology is classifying and identifying the disease categories [24]. Medical images are originated from various imaging techniques like CT scans, MRI, X-ray, microscopic, and ultrasound.

The major aim of the CAD system is to procure an agile and accurate diagnosis to assure the best treatment for a group of patients simultaneously. Furthermore, effective automated processing reduces the manual errors caused by humans, even processing time and cost is reduced [25].

A. **Image Acquisition:** Lung Dataset comprises of 267 lung CT scans and its respective mask scans [26]. This acquired dataset is divided into two partitions i.e. training and testing set. CT scans are grayscale images and size 128×128 pixels. Image preprocessing has been done to reduce the dimension by 32×32 to facilitate training easily. 128×128 dimensions have been resized to 32×32 to ease the training on low compute devices. Normalization is carried out on the entire dataset as shown in Figure 2.2 displays one of the sample scans on the left side whereas the right-hand side image is its segmented image.

B. **Image Preprocessing:** In preprocessing step images contrast has been enhanced by applying the Histogram Equalization (HE) method [44]. Different data augmentation techniques are rotations, zooming, shifting, cropping, and flipping, in the proposed methodology rotation technique is been applied for the data augmentation task. Figure 2.2 displays "8" distinct variants of CT scans obtained after applying rotation.

For image preprocessing, the HE (Histogram Equalization) technique has been applied to tune the intensity of the image dataset. The mathematical formula for tuning the image intensity is stated in Equation (2.1).

$$x^t = T(x) = \sum_{i=0}^{x} n_i \frac{max.intensity}{N} \tag{2.1}$$

Where, n_i represents no. of pixels at an intensity "I",
N represents total no. of image pixels.

This process improves the image contrast by portioning the image into n number of subparts and then it transforms the intensity value of all the subparts to achieve the appropriate histogram equalization concept. The following Figure 2.1 represents the enhanced image.

Afterimage enhancement data augmentation is applied by employing rotation operation on enhanced images. Position augmentation (i.e. altering the positions of image pixels) consists of various position altering methods like scaling, rotation, flipping, cropping, padding, translation, and affine transformation among these techniques rotation operation is performed.

Rotation image augmentation arbitrarily rotates the image in a clockwise direction by specified degrees ranging from 0 to 360. It will rotate the image pixels outside of the image frame by leaving the frame area empty i.e. pixel data is not accommodated in those areas. It's performed by using the following Keras package.

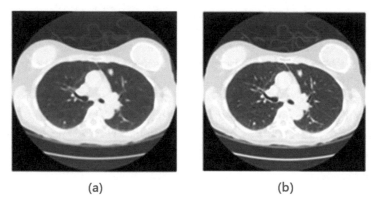

(a)   (b)

Figure 2.1   Representation of histogram equalization a: input image, and b: enhanced image.

$$\text{ImageDataGenerator}(\text{rotation\_range} = 10)$$

Where rotation_range = 10, implies that the angle to be rotated is "10'.

After the data augmentation step noise of the images is removed. There are various sources for noise in the images, which arrives from different aspects like image acquisition, image compression, and image transmission and even type of noises are different like Gaussian noise, Poisson noise, salt & pepper, speckle noise and so on. They are various image pre-processing algorithms for different categories of noises. Generally, they are three fundamental denoising techniques i.e. spatial filtering, domain transform filtering, and (the most adapted technique) wavelet thresholding. The key factor of filtering techniques while image denoising is: to preserve the edge information and to produce a natural image.

We had applied ABF (Adaptive Bilateral Filter) to sharpen the image and to eradicate noise from images as the two common image degradation problems are sharpness loss and noisy image. ABF nonlinear filter that smoothes the image and removes the noise and even preserves edge information. The bilateral filter shift-variant mathematical function is as explained in Equation (2.2)

$$f^{\wedge}(q, r) = \sum_{k} \sum_{l} h(q, r; s, t) g(s, t) \qquad (2.2)$$

Where, where $f^{\wedge}(q, r)$ represents image restored, and $g(q, r)$ represents degraded image, and $h(q, r; s, t)$ is the response of (q, r) to impulse of (s, t) [45].

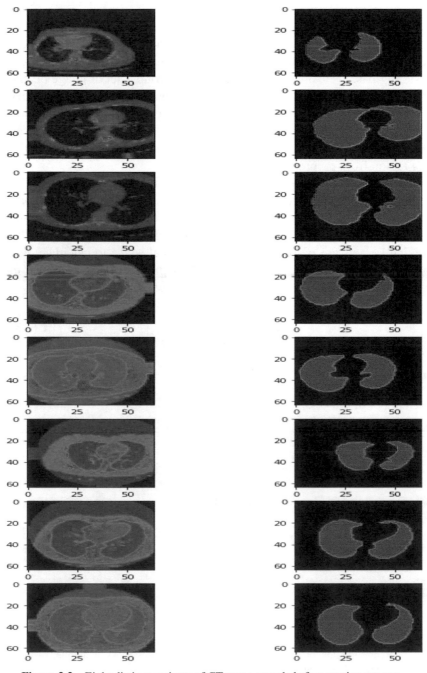

Figure 2.2 Eight distinct variants of CT scans recorded after rotation process.

C. Segmentation: it's the process of partitioning the image into various segments which are called image objects. Segmentation is performed to reduce the image complexity to simplify image analysis [46]. Various segmentation techniques like threshold method, region-based, edge-based, clustering-based, etc are used to partition and group a set of image pixels; labels are allocated in this process. These labels are used to identify boundaries, lines, and to separate necessary image objects i.e. ROI (region of interest) [47, 48]. In the proposed model segmentation is the key concept as it segments the region of interest i.e. cancer region from healthy lung region from preprocessed images.

Fuzzy C-Means (FCM) segmentation technique is employed. FCM method uses fuzzy partition technique to distinguish given data into 0 or 1 which determines the group. It partitions the finite set of elements let it be $X = \{0, 1, \dots, \}$ into a collection of C-fuzzy clusters. Acquired data by utilizing kernels are passed to multi-dimensional space to calculate data points. The mathematical function of FCM segmentation is as shown in Equation (2.3) and the segmented image is shown in Figure 2.3

$$f_m = \sum_{i=1}^{N} \sum_{j=1}^{C} u_{ij}^m |x_i - c_j|^2 \qquad (2.3)$$

Where, m = real no > 1
N = number of data,
C = number of clusters,
U_{ij} = degree of membership
x_i = ith data of d-dimension,
c_j = center of a cluster

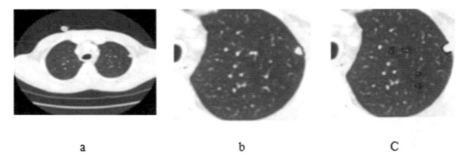

a   b   c

Figure 2.3   Representation of segmented results after applying FCM technique, a- input image, b- ground truth image, and c- segmented image.

D. Feature extraction: is the foremost step in the image processing method which employs various approaches to segment desired regions or features, and it's also useful for detecting a particular object in the image.

In the proposed method Grey Level Co-occurrence Method (GLCM) technique is employed for feature extraction. Various features like contrast, entropy, homogeneity, and maximum probability are extracted using this method. GLCM demonstrates the occurrence of various combinations of grey levels [51]. The normalization formula of GLCM is as shown in Equation (2.4)

$$P_{i,j} = \frac{V_{i,j}}{\sum\limits_{i,j=0}^{N-1} V_{i,j}} \tag{2.4}$$

Where i and j represent the row and column number

E. Feature Selection: Accurate and enhanced classification can be obtained by applying several features like texture, volumetric, and geometric features. As different features include a different kind of information, thus it results in an enhanced classification model by distinctive feature selection from different feature planes. The key concept of applying feature selection to the model is that it specifies and identifies the significance of the input feature set. Feature selection in our model is done by applying a Genetic Algorithm (GA).

GA is an empirical search approach that can be employed to find out or to search for optimal solutions [49, 50]. It can be applied and works well for unconstrained and unconstrained optimization challenges which are oriented upon natural selection method like methods that works on genetic evolvement. It consists of five stages viz initial population, fitness function, selection, crossover, and mutation. It is an iterative approach that comprises of population interacting with look space to find out the solution to the problems from a finite number of images known as "genome". Initial GA will continue as explores; the fundamental chromosome population is generated empirically. At every generation chromosomes are decrypted and evaluated in the second stage of GA viz. fitness function and the subsequent generation chromosomes are selected by a fitness function. Crossover and mutation stages produce new chromosomes in the population. This complete process is iterated till given fitness quality or maximum iterations are achieved. Feature selection optimal solution is determined by the best chromosome of the final generation.

Fitness function is described in Equation (2.5)

$$f(x) = \frac{f(i)}{f(I)}$$
(2.5)

Where *f(i)* determines individual fitness, and
F(I) determines sum of fitness of all individuals

Crossover mathematical formula is determined as shown in an Equation in (2.6)

$$C = \frac{(G + 2\sqrt{g})}{3G}$$
(2.6)

Where G determines the total number of evolutionary generated set by population, g determines the number of generations.

F. **Model Creation:** ConvNet and UNet architecture consists of convolutional layers ordered as top-down & bottom-up fashion shaping into u-shaped architecture. Thus, known as UNet model. Top-down route is the contracting pathway and bottom-up route is called the expansive pathway. Information of the image is acquired at the contraction stage and locating ROI is done at the expansion stage. Both the pathways consist of 4 blocks, where each block includes different layers like convolutional layers, ReLU function, and pooling layers followed by each other.

G. **Model Configuration:** Total consists of 32 layers where 1 input layer, 19 are convolutional layers, 4 max-polling layers, 4 up-sampling layers, and 4 concatenation layers.

Architecture: according to the U-Net, ResNet, and RCNN models we had implemented a Recurrent Residual CNN (RRCNN) for segmenting lung nodules. The proposed model makes use of above mentioned three model's properties. These three models are considered as they had performed well on medical image segmentation tasks. Proposed model performance can be evaluated mathematically according to RCL operations are carried out discretely [37–41].

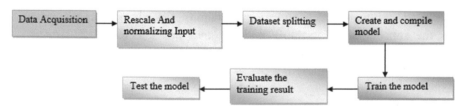

Figure 2.4   Architecture of Proposed Model.

Let's consider x_l as input to the lth a layer of RRCNN block and (i, j) as the center of pixel for the lesion patch of the input on the kth feature map in the RCL. And consider O_{ijk}^l as network output at t time, it is mathematically expressed as follows.

$$O_{ijk}^l(t) = (W_f^k)^T \times x_l^{f(i,\,j)}(t) + (W_r^k)^T \times x_l^{r(i,\,j)}(t-1) + b_k \quad (2.7)$$

Where, $(W_f^k)^T$ And $(W_r^k)^T$ are weights of standard convolution layer and k^{th} feature map of RCL respectively,
Convolution layer inputs are $x_l^{f(i,\,j)}(t)$ And $x_l^{r(i,\,j)}(t-1)$ and to l^{th} a layer of RCL, b_k represnsts bias.

RCL output is fed as input to activation function (ReLU) "f" and mathematically it is defined as

$$F(x_l,\ w_l) = f\left(O_{ijk}^l(t)\right) = \max(\,0,\ O_{ijk}^l(t)) \quad (2.8)$$

Where,
$F(x_l,\ w_l)$ is the output of the RCNN l^{th} layer,

This output of $F(x_l,\ w_l)$ is used in the convolutional encoding unit, decoding unit, and RuNet model respectively for upsampling and downsampling. Whereas in R2U-Net,

RCNN final outputs are passed to residual units referred in Figure 2.4.

Let's assume output of the RRCNN-block as, which is mathematically expressed as follows.

$$x_l + 1 = x_l + F(x_l, w_l) \quad (2.9)$$

Where, x_l is the input of RRCNN block
$x_l + 1$ is input to upsampling layers

However, the dimensions and number of feature maps for residual block and RRCNN block are the same as shown in Figure 1(d).

They are three approaches for feature selection i.e., forward selection, backward elimination, a combination of both. In the proposed approach we used information gain as a feature selection measure to enhance accuracy.

The proposed DL-based model is the building block of convolutional layers as referred to in Figures 2.4W(b) and (d). The U-Net model along with residual connectivity and recurrent convolution layers is implemented namely RRCNN as shown in Figure 2.4(d). Concatenation of feature maps for the model RRCNN is implemented from encoding to decoding unit [28, 29]. Advantages of proposed model architecture compared to U-Net i.e.,

Network parameters achieved high efficiency. i.e., the proposed model is designed in such a way that when compared to base models both include equivalent network parameters, and the proposed model achieved high performance for segmenting the lung nodules.

2.4 Experimental Analysis

Dataset: LUNA-16 (Lung Nodule Analysis) contest of Kaggle Data Science Bowl in 2017 had been conducted to detect lung lesions from both 2D & 3D lung CT scans. It comprises 267 2D CT scans and their relevant mask scans [20]. 512×512 is the size of each image which is reduced to 32×32 pixels. Python IDLE tool is used for evaluating experimental analysis.

Various performance measures like Accuracy (AC) and Dice coefficient (DC) are used to evaluate the proposed method.

To do this we also use the variables True Positive (TP), False Positive (FP), True Negative (TN), and False Negative (FN).

Accuracy is calculated using Equation (2.10).

$$AC = \frac{TP + TN}{TP + TN + FP + FN} \qquad (2.10)$$

Table 2.2   Demonstration of Confusion Matrix

	Predicted Patient with the Disease	Predicted Patient without Disease
Predicted with disease	TN	FN
Healthy persons	FP	TP

Table 2.3   Demonstrating dice-coef performance measure for various values of datasets i.e., 20%, 30% and 50% respectively as shown in below table

Epoch		1	2	3	4	5	6	7	8	9	10
Dice-Coef	Test_size= 20%	0.9215	0.9275	0.9405	0.9250	0.9388	0.9363	0.9347	0.9393	0.9415	0.9466
	Test_size= 30%	0.8848	0.887	0.9134	0.9244	0.9250	0.9234	0.9282	0.9312	0.9302	0.9312
	Test_size= 50%	0.8968	0.9295	0.9308	0.9342	0.9354	0.9359	0.9384	0.9395	0.9400	0.9400

Table 2.4 The following table illustrates various measures accuracy, precision, recall, and confusion matrix computed for various values of test size

Test Size	Accuracy	Precision	Recall	Confusion Matrix
80%–20%	97	92	97	[[163513 1755] [4229 51687]]
70%–30%	96	96	87	[[246463 10633] [3160 71520]]
50%–50%	97	96	92	[[412016 10894] [4901 121053]]

The DC is expressed as in Equation (2.11) where GT is Ground Truth and SR is Segmentation Result as shown in Figure 2.1.

$$DC = \frac{|GT \cap SR|}{|GT| + |SR|} \tag{2.11}$$

Graphs 1(a), (c), (e): represents training and validation accuracy. The X-axis denotes the number of epochs and the Y-axis specifies the accuracy obtained for each epoch respectively. Accuracy recorded for both training and validation is 96.78% & 96.98% respectively. Graph 1(b), (d), (f) exemplifies training loss and validation loss for various values of test data size where the X-axis indicates the epochs and the y-axis denotes cross entropy loss. Training loss obtained is 87.1% and 95.59% validation loss.

2.5 Cross Validation

Machine learning and deep learning models usually break down when induced on a dataset that has not been trained yet. In some cases, it works better and achieves the best accuracy and in some cases, it fails inadequately. To overcome this problem and to assure that the model can achieve the best results even on untrained data, we apply a re-sampling technique, called Cross-Validation. It is an estimation approach. Various ML, DL, and other models can be evaluated using this approach. Validation is carried out on subsets of testing data. The overfitting problem can be identified by the cross-validation technique [52].

It is carried out in three steps:

i Keep aside some part of the sample test dataset.
ii Train the model on the rest of the train data.
iii Know to utilize the reserved part of the dataset to test the model.

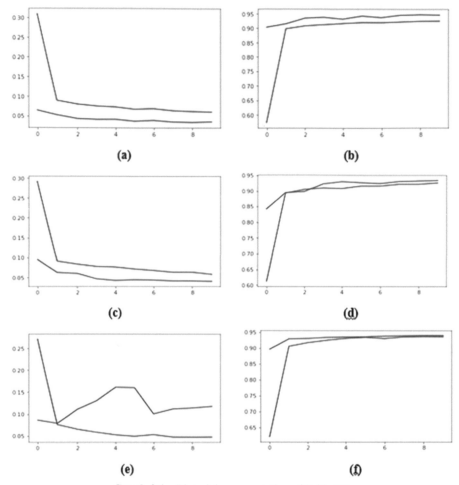

Graph 2.1   Pictorial representation of Table 2.3.

Usually, a simple technique is used for this purpose i.e., dividing the dataset into 3 fractions i.e., training, testing, and validation set. But the problem with this technique is that it doesn't work well on the small dataset as splitting the small dataset may result in the loss of some useful information from the training procedure and this causes the model to break down. There are different types of cross-validation methods and broadly categorized into two types i.e., Exhaustive and Non-exhaustive cross-validation. We had applied k fold cross-validation technique.

Table 2.5 Performance measure of K-fold cross-validation

S.No	K-value	Accuracy Obtained
1	$K=1$	92.23
2	$K=2$	90.53
3	$K=3$	92.20
4	$K=4$	92.30
5	$K=5$	92.06

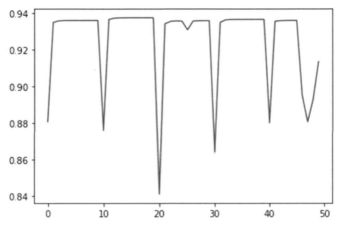

Graph 2.2 Cross-validation result.

Working of K-fold Cross-validation: The resampling method utilized for estimating ML and DL models is known as Cross-Validation [53]. First, the dataset is divided randomly into 'k' number of equal subsets, and then one of the subsets is considered for the evolution dataset and the rest of the subsets are considered as testing dataset. This process continues until all the k subsets are evaluated [54]. The primary advantage of this method is all deliberations are employed while both training and validating the network.

2.6 Conclusion

In the present study we implemented the Recurrent Residual Convolutional Neural Network paradigm the enhanced model of UNet using recurrent network and residual networks. The proposed model is evaluated on LUNA16 for lung nodule classification and prediction. The outcome illustrates that the proposed model shows enhanced performance in segmentation tasks in contrast to existing techniques i.e., UNet and ResNet model. Also, the result demonstrates proposed system not just provides enhanced performance at training the model, but the same performance is provided even at testing and validation. 97% accuracy is achieved, 92% precision and 97% recall. Further, the system can be improved by using the actual size of an input image, and a total number of layers can be increased which enhances the accuracy.

References

[1] Overman, V. P., American Cancer Society[J]. International Journal of Dental Hygiene, 2006, 4(2):109–109.

[2] https://theappsolutions.com/blog/development/machine-learning-algorithm-types/

[3] Lung Cancer Symptoms, Types, Causes, Treatment & Diagnosis, https://www.medicinenet.com/lung_cancer/article.html.

[4] Bhatia, S., Sinha, Y., & Goel, L. (2018). Lung Cancer Detection: A Deep Learning Approach. Soft Computing for Problem Solving, 699–705. doi:10.1007/978-981-13-1595-4_55.

[5] A. Asuntha, A. Brindha, S. Indirani, Andy Srinivasan, "Lung cancer detection using SVM algorithm and optimization techniques", Journal of Chemical and Pharmaceutical Sciences, ISSN: 0974–2115.

[6] Fangzhou Liao, Ming Liang, Zhe Li, Xiaolin Hu, (2019), "Evaluate the Malignancy of Pulmonary Nodules Using the 3-D Deep Leaky Noisy-or Network", IEEE Transactions on Neural Networks and Learning Systems, 1–12. doi:10.1109/tnnls.2019.2892409.

[7] QingZeng Song, Lei Zhao, XingKe Luo, and XueChen Dou, (2017), "Using Deep Learning for Classification of Lung Nodules on Computed Tomography Images", Journal of Healthcare Engineering Volume 2017, https://doi.org/10.1155/2017/8314740.

[8] Md Zahangir Alom, Mahmudul Hasan, Chris Yakopcic, Tarek M. Taha, Vijayan K. Asari, (2018), "Recurrent Residual Convolutional Neural Network based on U-Net (R2U-Net) for Medical Image Segmentation", arXiv:1802.06955.

[9] Petros-Pavlos Ypsilantis Giovanni Montana, (2016), "Recurrent Convolutional Networks for Pulmonary Nodule Detection in CT Imaging", arXiv:1609.09143.

[10] Lakshmanaprabu S.K., Sachi Nandan Mohanty, Shankar K., Arunkumar N., Gustavo Ramirez, (2018), "Optimal deep learning model for classification of lung cancer on CT images", Future Generation Computer Systems (2018), https://doi.org/10.1016/j.future.2018.10.009.

[11] A.Worawate. Thirach, A., Marukatat, S., & Wilaiprasitporn, T, (2018), "Automatic Lung Cancer Prediction for Chest X-ray Images Using the Deep Learning Approach", Biomedical International Conference (BMEiCON-2018).

[12] Jason L. Causey, Yuanfang Guan, Wei Dong, Karl Walker, Jake A. Qualls, Fred Prior, Xiuzhen Huang, (2019), "Lung Cancer Screening With Low-Dose CT Scans Using A Deep Learning Approach.", ArXiv:1906.00240 [eess.IV].

[13] Christoph Haarburger, Philippe Weitz, Oliver Rippel, Dorit Merhof, (2019), "Image-Based Survival Prediction for Lung Cancer Patients Using CNN", 2019 IEEE 16th International Symposium on Biomedical Imaging (ISBI 2019), doi:10.1109/isbi.2019.8759499.

[14] Ahmed Elnakib, Hanan M. Amer, Fatma E.Z. Abou-Chadi, (2020), "Early Lung Cancer Detection using Deep Learning Optimization", *International Journal Of Online and Biomedical Engineering (IJOE)*, Vol. 16.

[15] Ait Skourt Brahim., El Hassani, A., & Majda, A. (2018), "Lung CT Image Segmentation Using Deep Neural Networks", Procedia Computer Science, 127, 109–113. doi:10.1016/j.procs.2018.01.104.

[16] Atsushi Teramoto, Tetsuya Tsukamoto, Yuka Kiriyama, and Hiroshi Fujita, "Automated Classification of Lung Cancer Types from Cytological Images Using Deep Convolutional Neural Networks", BioMed Research International Volume 2017, Article ID 4067832, 6 pages https://doi.org/10.1155/2017/4067832.

[17] Kuruvilla, J., & Gunavathi, K, (2014), "Lung cancer classification using neural networks for CT images". Computer Methods and Programs in Biomedicine, 113(1), 202–209. doi:10.1016/j.cmpb.2013.10.011.

[18] Hu, Z., Tang, J., Wang, Z., Zhang, K., Zhang, L., & Sun, Q. (2018). Deep learning for image-based cancer detection and diagnosis-A survey. Pattern Recognition, 83, 134–149. doi:10.1016/j.patcog.2018.05.014.

[19] Chougrad, H., Zouaki, H., & Alheyane, O, (2019), "Multi-label Transfer Learning for the Early Diagnosis of Breast Cancer", Neurocomputing, doi:10.1016/j.neucom.2019.01.112.

[20] Cascio, D., Magro, R., Fauci, F., Iacomi, M., & Raso, G, (2012), "Automatic detection of lung nodules in CT datasets based on stable 3D mass–spring models", Computers in Biology and Medicine, 42(11), 1098–1109. doi:10.1016/j.compbiomed.2012.09.002

[21] Fausto Milletari, Nassir Navab, "V-Net: Fully Convolutional Neural Networks for Volumetric Medical Image Segmentation", 2016 Fourth International Conference on 3D Vision.

[22] Zifeng Wu, Chunhua Shen, Anton van den Hengel (2019), "Wider or Deeper: Revisiting the ResNet Model for Visual Recognition", doi:10.1016/j.patcog.2019.01.006.

[23] Olaf Ronneberger, Philipp Fischer, and Thomas Brox, (2015), "U-Net: Convolutional Networks for Biomedical Image Segmentation". Medical Image Computing and Computer-Assisted Intervention – MICCAI 2015, 234–241. doi:10.1007/978-3-319-24574-4_28.

[24] Ahmed Elnakib, Hanan M. Amer, Fatma E.Z. Abou-Chadi, (2020), "Early Lung Cancer Detection using Deep Learning Optimization", International Journal Of Online and Biomedical Engineering (IJOE), Vol. 16.

[25] Ait Skourt Brahim., El Hassani, A., & Majda, A. (2018), "Lung CT Image Segmentation Using Deep Neural Networks", Procedia Computer Science, 127, 109–113. doi:10.1016/j.procs.2018.01.104.

[26] Atsushi Teramoto, Tetsuya Tsukamoto, Yuka Kiriyama, and Hiroshi Fujita, "Automated Classification of Lung Cancer Types from Cytological Images Using Deep Convolutional Neural Networks", BioMed Research International Volume 2017, Article ID 4067832, 6 pages https://doi.org/10.1155/2017/4067832.

[27] Kuruvilla, J., & Gunavathi, K, (2014), "Lung cancer classification using neural networks for CT images". Computer Methods and Programs in Biomedicine, 113(1), 202–209. doi:10.1016/j.cmpb.2013.10.011.

[28] Hu, Z., Tang, J., Wang, Z., Zhang, K., Zhang, L., & Sun, Q. (2018). Deep learning for image-based cancer detection and diagnosis-A survey. Pattern Recognition, 83, 134–149. doi:10.1016/j.patcog.2018.05.014.

[29] Chougrad, H., Zouaki, H., & Alheyane, O, (2019), "Multi-label Transfer Learning for the Early Diagnosis of Breast Cancer", Neurocomputing, doi:10.1016/j.neucom.2019.01.112.

[30] Cascio, D., Magro, R., Fauci, F., Iacomi, M., & Raso, G, (2012), "Automatic detection of lung nodules in CT datasets based on stable 3D mass–spring models", Computers in Biology and Medicine, 42(11), 1098–1109. doi:10.1016/j.compbiomed.2012.09.002.

[31] Fausto Milletari, Nassir Navab, "V-Net: Fully Convolutional Neural Networks for Volumetric Medical Image Segmentation", 2016 Fourth International Conference on 3D Vision.

[32] Zifeng Wu, Chunhua Shen, Anton van den Hengel (2019), "Wider or Deeper: Revisiting the ResNet Model for Visual Recognition", doi:10.1016/j.patcog.2019.01.006.

[33] Olaf Ronneberger, Philipp Fischer, and Thomas Brox, (2015), "U-Net: Convolutional Networks for Biomedical Image Segmentation". Medical Image Computing and Computer-Assisted Intervention – MICCAI 2015, 234–241. doi:10.1007/978-3-319-24574-4_28.

[34] Jump, Schmidhuber, J. (2015). "Deep Learning in Neural Networks: An Overview". Neural Networks. 61: 85–117.

[35] Litjens, Geert, Thijs Kooi, Babak Ehteshami Bejnordi, Arnaud Arindra Adiyoso Setio, Francesco Ciompi, Mohsen Ghafoorian, Jeroen AWM van der Laak, Bram van Ginneken, and Clara I. Sánchez. "A survey on deep learning in medical image analysis." Medical image analysis 42 (2017): 60–88.

[36] Han, Zhongyi, Benzheng Wei, Yuanjie Zheng, Yilong Yin, Kejian Li, and Shuo Li. "Breast cancer multi-classification from histopathological images with structured deep learning model." Scientific reports 7, no. 1 (2017): 4172.

[37] Han, Zhongyi, Benzheng Wei, Yuanjie Zheng, Yilong Yin, Kejian Li, and Shuo Li. "Breast cancer multi-classification from histopathological images with structured deep learning model." Scientific reports 7, no. 1 (2017): 4172.

[38] Greenspan, Hayit, Bram van Ginneken, and Ronald M. Summers. "Guest editorial deep learning in medical imaging: Overview and future promise of an exciting new technique." IEEE Transactions on Medical Imaging 35, no. 5 (2016): 1153–1159.

[39] Larsson, Gustav, Michael Maire, and Gregory Shakhnarovich. "FractalNet: Ultra-Deep Neural Networks without Residuals." arXiv preprint arXiv:1605.07648 (2016).

[40] Zeiler, M. D. and Fergus, R. Visualizing and understanding convolutional networks. CoRR, abs/1311.2901, 2013. Published in Proc. ECCV, 2014.

[41] Szegedy, Christian, et al. "Going deeper with convolutions." Proceedings of the IEEE conference on computer vision and pattern recognition. 2015.

[42] Simonyan, Karen, and Andrew Zisserman. "Very deep convolutional networks for largescale image recognition." arXiv preprint arXiv:1409.1556(2014).

[43] Esteva, Andre, Brett Kuprel, Roberto A. Novoa, Justin Ko, Susan M. Swetter, Helen M. Blau, and Sebastian Thrun. "Dermatologist-level classification of skin cancer with deep neural networks." Nature 542, no. 7639 (2017): 115.

[44] Ani Brown Mary N, Dharma D (2017) "Coral reef image classification employing improved LDP for feature extraction", Elsevier. J Vis Commun Image Represent 49(C):225–242. https://doi.org/10.1016/j.jvcir.20 17.09.008

[45] Zhang B, Allebach JP (2008) Adaptive Bilateral Filter for Sharpness Enhancement and Noise Removal. IEEE Trans Image Process 17(5).

[46] Ait Skourt, B., El Hassani, A., & Majda, A. (2018). Lung CT Image Segmentation Using Deep Neural Networks. Procedia Computer Science, 127, 109–113. doi:10.1016/j.procs.2018.01.104

[47] K. Nakagomi, A. Shimizu, H. Kobatake, M. Yakami, K. Fujimoto, and K. Togashi, "Multi-shape graph cuts with neighbor prior constraints and its application to lung segmentation from a chest CT volume," Med. Image Anal., vol. 17, no. 1, pp. 62–77, Jan. 2013.

[48] S. Hu, E. A. Hoffman, and J. M. Reinhardt, "Automatic lung segmentation for accurate quantitation of volumetric X-ray CT images," IEEE Trans. Med. Imaging, vol. 20, no. 6, pp. 490–498, Jun. 2001.

[49] Lu, C., Zhu, Z., & Gu, X. (2014). An Intelligent System for Lung Cancer Diagnosis Using a New Genetic Algorithm Based Feature Selection Method. Journal of Medical Systems, 38(9). doi:10.1007/s10916-014-0097-y

[50] Liang, C., and Peng, L., An automated diagnosis system of liver disease using artificial immune and genetic algorithms. J. Med. Syst. 37(2):1–10, 2013.

[51] Adi, Kusworo; Widodo, Catur Edi; Widodo, Aris Puji; Gernowo, Rahmat; Pamungkas, Adi; Syifa, Rizky Ayomi. (2018). Detection Lung Cancer Using Gray Level Co-Occurrence Matrix (GLCM) and Back Propagation Neural Network Classification. Journal of Engineering Science & Technology Review. 2018, Vol. 11 Issue 2, p. 8–12. 5p.

[52] Max A. Little, Gael Varoquaux, Sohrab Saeb, Luca Lonini, Arun Jayaraman, David C. Mohr and Konrad P. Kording. (2017). Using and understanding cross-validation strategies. Perspectives on Saeb et al. GigaScience, 6, 2017, 1–6

[53] https://www.digitalvidya.com/blog/cross-validation-in-machine-learni ng/

[54] https://machinelearningmastery.com/k-fold-cross-validation/

3

Machine Learning Application for Detecting Leaf Diseases with Image Processing Schemes

Canavoy Narahari Sujatha[1] and V. Padmavathi[2]

[1]Sreenidhi Institute of Science and Technology, India
[2]Anurag Univerity, India
cnsujatha@gmail.com; chpadmareddy1@gmail.com

Abstract

This chapter reports on the evolution and performance of machine learning with image processing in detecting plant leaf diseases. Plant leaf diseases and harmful insects pose a significant threat to agriculture. A more reliable and faster prediction of leaf diseases in crops could aid in the development of an early treatment method. Modern advanced developments in Machine Learning and its advancements have allowed to extremely improve the performance and accuracy of object detection and recognition systems. Identification of plant diseases using supervised learning and control of plant diseases, detecting nutrient deficiency, controlled irrigation, and controlled use of fertilizers and pesticides are all part of crop management from the early stages to the mature harvest period. If certain diseases are identified in the initial stage can be cured and the proper pesticide usage can help in the proper plant growth. This cannot be done easily efficiently even by the so experienced farmers, so 225-HMP60A moisture sensor is used for monitoring the moisture levels in the soil because having high moisture makes it vulnerable to pest attack.

Keywords: Machine Learning, Leaf diseases, Segmentation, Convolutional Neual Network, Precision, Recall, f1-score.

3.1 Introduction

Since vision is the most important sense in humans, images have played an important role in their lives. As a result, the field of image processing has a wide range of applications. Images are everywhere nowadays, more than ever, and thanks to developments in digital technology, it is very simple for anyone to produce a large number of images. Traditional image processing methods must deal with more complex problems as a result of the abundance of images, as well as their adaptability to human vision. Then machine learning has developed as a solution to the complexity of vision. Machine learning with image processing has newly gotten a lot of attention thanks to the introduction of image datasets and benchmarks. A groundbreaking application of machine learning to image processing would almost certainly be beneficial to the industry, allowing for a deeper understanding of complex images. As the need for adaptation grows, the number of image processing algorithms that integrate certain learning components is expected to grow. However, an increase in adaptation is often associated with an increase in complexity, and any machine learning method must be effectively controlled in order to be properly adapted to image processing problems. The processing of large images necessitates the ability to process a large amount of data, which is troublesome for most machine learning techniques. Interaction with image data and image priors is needed to drive model selection strategies.

Nowadays, Agriculture is the science and the art of cultivating. Cultivation is a tedious work. After all the efforts crop does not grow due to an inadequate amount of moisture in the soil and infections. Crop loss is 20 to 40 percent caused by infection, weed, and animals. India has a crop loss of 15 to 20 percent by Proximal Convoluted Tubule. Now by using advanced technology like deep learning it became tranquil to identify the infection and level of moisture for better production of a healthy crops.

This proposed idea can help to identify the infection and control it thereby reducing crop loss. This is the main objective of the chapter. Machine learning is implemented by the use of libraries of python and supervised learning. A wide range of data sets of 87867 images is used for training the machine. These images are subdivided into train data, valid data, and test data.

Now the output of the test image after classifying and identifying with name of the disease. Outputs as precision which states the accuracy or how sure the machine is about the output. It can be increased by multiplying the images used for the learning process i.e. train dataset. Plant diseases pose a significant threat to food security, but in many parts of the world, rapid isolation is difficult due to a lack of an essential base. The evolution of accurate techniques in the leaf-based imaging field has shown surprising outcomes. Using machine learning algorithms to train and predict, publicly available data sets gives a clear pathway to identify the infection present in plants.

An expert in the agricultural regions of the provinces may think that it is difficult to distinguish potential diseases from their harvest. It is not possible to reach out to the agribusiness office every time and find out which infection it is. Our goal is to distinguish a plant-derived illness by looking at its morphology through image processing and machine learning. Diseases from pests result in the loss of crops or a part of crop cumulating in a reduction in food production will lead to food insecurity. Information on infection administration or control and disease is limited in various advanced countries. Poisonous germs, destitute disease control, the severe temperature change are just some of the key factors in producing reduced food. Distinct contemporary technologies have been identified to reduce the postharvest process, enhance cultivation stability and increase productivity. To recognize diseases, a variety of laboratory techniques have been used, including gas chromatography, mass spectrometry, thermography, polymerase ketone reaction, and hyper spectroscopy. These processes, on the other hand, are both inexpensive and time-consuming. To diagnose diseases, server-based and cell-based diagnostic approaches have recently been used.

A few of these methods being high-performance processing, high resolution camera, and many extensions are designed to add additional benefits that lead to automatic infection recognition. Modern methods like machine learning using deep learning algorithms are deployed to increase the amount of recognition and accuracy of outcomes. Various studies have been conducted under the plant pathology field to study Whole forests, classification methodology, recording, and other works that work on building forests of decision trees during training. Unlike decision trees, Random forests overcome the disadvantages of the most relevant training data to handle both categorical and statistical information. The histogram of trend gradients (HOG) is a description used for image processing and PC vision due to object detection. SVM works by converting samples into points in feature space, where samples of different classes can be separated on opposite sides of an optimal hyperplane. SVM can perform both linear and nonlinear classifications using a set of mathematical functions that transform raw data to required forms for better feature extraction. These functions are called kernel functions. the histogram of oriented gradients (HOG) image feature descriptor is combined with SVM for better feature extraction. In HOG detection, the images are broken down into connected regions called blocks (for example 12×12 pixels per block). Each block is divided into sub-cells. Spatial gradients of pixel values are computed for each pixel in the cells. Each cell is discretized into angular bins based on the gradient orientation. Lastly, histograms of the gradients in each block are computed and serve as the feature descriptors.

In machine learning, higher-dimensional feature space is more likely to yield more separable features. This could be accomplished by using general functions such as polynomial and radial basis functions to transform features (RBF). Transforming features may result in a substantial increase in feature dimensions, and thus an increase in training time. By measuring the dot products of the features rather than actually converting them, we could convert features to far higher dimensions using SVM. Decision trees (DT) classifiers are created by dividing data into smaller groups and determining which division produces the greatest disparity. For disparity measures, the Gini index or entropy are commonly used [17]. The benefit of DT is that the findings can be deciphered by humans. If we let the tree learn without any depth constraints, a DT could produce zero training errors. Ensemble methods are commonly used to avoid a tree being overfitted to the data. Random Forest is a common ensemble method for DT (RF). Random forests are made up of a number of trees, each of which is trained using a subset of the data and features. As a result, each tree can be thought of as a simpler tree. The decisions from all trees are averaged or based on a majority vote for class decisions given test results. The Naive Bayes (NB) classifier is a variant of the Bayes classifier in which each feature is assumed to be conditionally independent of the others. With this assumption, the bayes theorem can be used to calculate the posterior probability of data belonging to a class label using only the product of conditional probabilities for each feature. NB is very useful in many classification activities, despite the fact that this assumption does not hold in most real-world situations. Like that many traditional and advanced technologies are combined to select machine learning models to get an acceptable level of accuracy.

- Machine learning is alienated into four categories: supervised, semi-supervised, unsupervised, and reinforcement learning. It provides a training collection of instances with suitable goals to a computer system through supervised learning. Using this training set scheme, provide accurate responses to provided possible inputs. The two sub-categories of Supervised Learning are classification and regression.

- Using classification methods, the inputs are divided into different classes, and the trained system must produce behavior that assigns hidden inputs to these classes. This is referred to as the multi-labeling method. Spam purification is a classification process in which emails are labeled as "spam" or "not spam".
- The regression is a supervised technique with the continuous outcomes rather than discrete outcomes. Unlike classification predictions, where accuracy is used as a success metric, regression predictions are evaluated using root mean squared error (RMSE).
- Unsupervised learning entails the machine making its own decisions rather than relying on a dataset to practice. The machine receives no labeling that can be used to make forecasts. With the support of feature learning of the provided data, unsupervised learning can be used to retrieve the hidden pattern.
- Clustering is an unsupervised learning method for dividing inputs into groups. These clusters were not previously known. It creates groups based on resemblance.
- Semi-supervised learning - The scheme is believed to be partial training data in semi-supervised learning. This form of training is done with some previously trained data in order to find some missing results. This form of algorithm is used for training commitment on untagged data. The semi-supervised learning algorithm is trained on both labeled and unlabeled data, and it possesses the characteristics of both unsupervised and supervised learning algorithms.
- Reinforcement learning - The trained data is only given as an answer to the program's actions in a self-motivated situation, such as driving a car or playing a video game, in Reinforcement learning.

Machine learning algorithms are very effective in classifying plant leaves with specific diseases. Furthermore, deep learning methods are an advanced process of machine learning algorithms that use neural networks to identify data and predict more accurately. Machine Learning algorithms use a number of techniques to make decisions based on vast quantities of complex data. With unique inputs provided to the machine, these algorithms complete the task of learning from data. Input dataset – Preprocessing – Feature extraction – Model Creation – Training - Prediction – Analysis - Retraining – Prediction is the cycle of any ML Algorithm. This allows the machine learning algorithm to learn on its own and generate the best possible response, which will improve in accuracy over time. Figure 3.1 shows a block diagram of the phases of the machine Learning Model for image classification/identification.

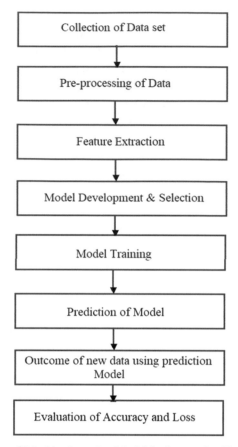

Figure 3.1   Phases of Machine Learning Model for image classification/identification.

Eliminating variable lighting effects, transferring the image to a different color space, choosing a suitable color channel, enhancing the selected color channel, contrast enhancement, normalising color variance induced by image acquisition, smoothing the image, and removing the vignetting effect are all basic pre-processing measures. a good combination of pre-processing steps will help a lot inaccurate image segmentation. Image segmentation methods include histogram thresholding, clustering, localized and distributed area recognition, active contours (snakes), edge detection, supervised learning, graph theory, and probabilistic modeling, among others. To achieve optimum precision, these methods can be used individually or in combination. For accurate image segmentation, feature detection, and generation, post-processing is needed.

Convolutional Networks are composed of multilayer neurons with learnable weights and biases. Neurons are mathematical models where weights, biases, and non-linear functions are applied to input data (images plus labels) for feature extraction. Random weights and biases are generated at the beginning of the computation. The values of weights and biases are optimized during model training and the optimized values can be used for prediction. There are three main kinds of structure in Convolutional Networks like convolutional layers, pooling layers, and fully connected layers as illustrated in Figure 3.2. Convolutional layers take filters of different sizes and slide them over the complete image. The dot products between the values in the filter and the pixel values of the input images are calculated. The results are passed to pooling layers for dimensionality reduction, where some results are dropped randomly. In the end, fully connected layers are used to perform image classification based on the results from pooling layers. Each layer is connected to another through a differentiable function. Softmax is one of the most widely used differentiable functions. In classification tasks, the softmax function calculates the probability (P) for the kth class (in our case, infected leaf images or non-infected images) given input vector X (in our case, processed image pixel values) and weighting vector W, which is learned by the model.

There are many tricks in Convolutional Networks including shared weights, down-sampling, and translation invariance that enable them to classify images efficiently. ConvNets are especially good at image processing

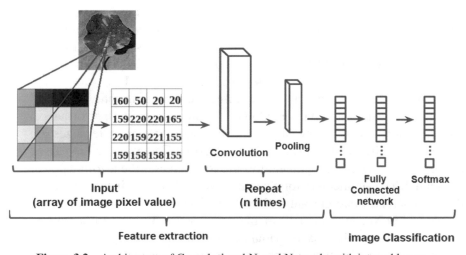

Figure 3.2 Architecture of Convolutional Neural Networks with internal layers.

tasks by preserving spatial information and reducing model parameters through weight sharing between the neurons. The Convolutional Networks can be built from scratch or be pretrained on large image datasets. The architecture of Convolutional Networks built from scratch is illustrated in Figure 3.4. Weight layers are where weights get trained and updated. Since pooling layers only perform dimension reduction on weights without training the weights, pooling layers are not counted as weight layers. There are 8 weight layers, including 6 convolutional layers and 2 fully connected layers. Sometimes, it is difficult to gather a dataset that is sufficiently representative of the real world. Furthermore, training Convolutional Networks from scratch can be time-consuming and tedious. To solve these problems, transfer learning algorithms are developed. In transfer learning, Convolutional Networks of various architectures are trained on a large dataset like ImageNet, which contains 1.2 million images divided into 1000 categories.

Once these Convolutional Networks models are tuned and the optimal structure and model parameters are obtained.

These models can be implemented as a fixed feature extractor to solve new problems. To get the best performance model, evaluation can be done on pre-trained Convolutional Network models for infected leaf image classification: vgg16, vgg19, xception, and inceptionv3. The pre-trained models run much faster than Convolutional Networks which are trained from scratch and often give more accurate prediction results.

Multiple 3×3 convolutional layers are adopted in the architecture of vgg networks to represent complex features. A massive number of parameters are created as the number of convolutional layers in a vgg network increases, and learning power is greatly increased. The vgg16 model has 16 weight layers, including 13 convolutional layers and three fully connected layers. The vgg19 model has 19 weight layers, including 16 convolutional layers and three fully-connected layers. Similarly, the architecture for the inceptionv3 model has a special inception module is adopted to improve model performance. In the inception module, convolutional layers with different filter sizes are computed in parallel and the resulting features are concatenated before passed to the next layer. The increase in features greatly increases the learning power of the model. The architecture for Xception improves on the inception model with a simple and more elegant architecture. 36 stages of convolutions are implemented in Xception models, including 34 separable convolutional layers. Inseperable convolutions, a spatial convolution is implemented keeping the channels separately, then a depth wise convolution is performed.

Table 3.1 Confusion Matrix

	Predicted Class	
Actual Class	Class = True	Class = False
Class = True	True Positive (TP)	False Negative (FN)
Class = False	False Positive (FP)	True Negative (TN)

Finally, to evaluate the performance of classifiers a confusion matrix is used. In a confusion matrix, the columns denote the predicted labels from the algorithm and the rows denote the true labels. A confusion matrix for a binary classification problem is shown in Table 3.1; this can be expanded to compute a confusion matrix for a multiclass classification problem.

The values from the confusion matrix are used to calculate the commonly used evaluation metrics such as Recall, precision, and Accuracy. Based on the observed accuracy values of the model, to maximize the accuracy the chosen model gets trained for more iterations with a large amount of image data either by collecting a large datasets or using the image augmentation technique.

3.2 Existing Work on Machine Learning with Image Processing

In this section for disease detection, some segmentation and extraction methods are used. Economic and social losses can be caused by many diseases that have an effect on plants. The loss to the product has been avoided by detecting the disease at the initial stage. Different existing methods and approaches are discussed in this paper. Initially, it is difficult to develop a single standard system for diagnosing all types of diseases. This paper proposes a limited disease type program, so in the future, the work should be expanded by designing different strategies for better diagnosis.

Jiang Lu, et_al [1] identified their project as an automated in-field diagnosis program of wheat disease, based on a weekly supervised framework of deep learning, which incorporates identification for diseases in wheat and localizes disease zones with only annotations of image level to wildlife-related training images. In addition, the Wheat Disease Database 2017 (WDD2017) new infield image data collection for wheat diseases is being collected to verify the system's performance. Under two architectures, i.e. The mean accuracies of VGG -FCNVD16 and VGG-FCN-S, which have been surpassed by 93.28 percent and 73.01 percent by two traditional CNN sets, are 97.94 percent and 95, 11 percent respectively for five folds over cross-validation on the WDD2017 VGG – CNN - VD16 and CNN - S - VGG The experimental results show that, while retaining accurate monitoring for the corresponding disease areas, the device proposed exceeds the traditional CNN architectures for recognition accuracy in a comparable amount. In addition, the program proposed for the treatment of agricultural diseases has been developed into a smartphone device for the first time.

A 40 research efforts using profound learning methods applied to various challenges in farming and food production were addressed and surveyed [2]. Examine the specific farming flaws under investigation, the various modeling and frameworks used, the data sources, data design and pre-processing, and the overall performance of the metrics used in each study. Furthermore, contrasts in terms of classification or regression performance of profound learning and other common techniques are being studied. Deep learning outperforms current image processing methods in terms of precision and accuracy, according to the findings.

In [3] models of the neural convolutional network for identification and diagnosis of diseases in plants with simple photographs of some healthy and some diseased plants through methods of deep learning. An open database of 87,849 images containing 25 different plants in a collection of 57 distinct groups of [plant, disease] combinations, including healthy plants, was used to train the model. The best model architecture achieved a 99.529 percent success rate in classifying/recognizing the corresponding [plant, disease, or healthy] combination. This model's high success rate makes it a very useful early warning or advisory tool, as well as a technique that can be used to support an integrated system of plant disease detection in working farm situations.

A methodology for the detection of crucial and precise plant leaf diseases with the help of artificial networks and various image processing techniques is defined [4]. Since the proposed approach is based on the ANN classifying and a Gabor feature removal filter, improved results are obtained with a recognition rate of capable of 91percent. To classify various plant leaf diseases and combine them with textures, color, and morphology, an ANN-based classifier is used.

Authors in [5] introduced an important approach like K-means- clustering, texture, and color analysis for the identification of diseases in Malus domestica. This uses the color and texture characteristics that normally occur in natural and affected zones to discriminate and recognize precise agriculture. In [6], the efficiency of traditional multiple analysis, artificial neural network (back propagation neural network) and supporting vector machine

(SVM) were examined by the authors. It was concluded that a statistical approach based on SVM directed to a better description of the connection between environmental circumstances and the extent of disease could be helpful for disease supervision.

In [7], the authors presented a review and diagnose the need to develop a speedy, cost-effective, and consistent health-monitoring sensor that enables progress in agriculture. They designated the technologies presently used, including spectroscopic, imaging based and volatile profiling based methods for the discovery of plant diseases, in order to develop a ground based sensor scheme to help spectate plant health and diseases under field situations/conditions. Upon reviewing the study and evaluating the findings of their work on papers [8–11], it has been determined that image processing approaches such as double-stranded ribonucleic acid (RNA) analytics, nucleic acid samples, and microscopy should also be applied to known diseases amongst additional approaches widely applied for plant disease diagnostics.

Many methods for the identification of plant diseases by computer vision are currently in use. One is the identification of diseases by removing color as described in [12]. This paper used coloring models for YcbCr, HSI, and CIELB, and thus the noise from different sources, for example camera flash, was observed in disease spots and not affected.

Furthermore, the method of extracting shape features may achieve plant disease detection. In sugarcane blades where threshold segmentation for determining the area of the leaf and the triangulate threshold for the lesion region was used by Patil and Bodhe for the detection of the disease, they obtained an average of 98.59% accuracy in the final experiments [13]. In addition, the texture extraction feature may be used in the detection of plant diseases. Patil and others proposed a plant disease detection model using texture characteristics such as homogeneity, inertia, and correlation by measuring the co-occurrence gray level matrix on the picture [14]. They experimented with the identification of diseases on maize leaves in conjunction with color extraction. The amalgamation of all these features delivers a comprehensive set of features for enhancing and classifying pictures. As submitted a survey in [15], well-known conventional methods for extraction of features are discussed. The surveyed research is primarily about pertaining to these methodologies and techniques, despite the quick advancement of an artificial intelligence sciences (AI).

There are several methods that narrate the feed-forward spread of neural networks containing one input, one output, and one hidden layer for an identification of leaf, disease, or pest species; this type of model was projected in paper [16]. They constructed a software model to recommend enforcement action in agricultural crops for pests or diseases. As suggested in [17] and [18] features extracted by PSO are included and forward neural networks for the determination of the damaged cotton leaf spot, which increase the device accuracy and eventually, the overall accuracy of 94 percent. In this regard, the authors proposed to apply the same technique.

Recognition and standardization of plant leaf diseases can also be achieved with support vector machine algorithms. This procedure was applied to sugar beet diseases and was introduced in [19]. But depending on the type and phase of the disease, the accuracy of categorization was between 66 and 90 percent. Similarly, there are methods that combine the extraction feature and the Neural Network Ensemble (NNE) for the recognition of plant leaf disease. Through the training of numerous neural networks and the combination of their results, Neural Network Ensemble suggests a well generalization of learning skills [20]. Such a method was applied only for the recognition of tea leaf diseases with a final test accuracy of 91 percent [21].

A further tactic based on leaf images and with ANNs as a system for the instinctive detection and categorization of plant diseases was used in combination with means as a clustering procedure proposed in [22]. ANN was made up of ten hidden layers. The number of outputs was six, which corresponded to the number of groups that represented five diseases and a healthy leaf. The accuracy of classifications using this method was 94.65% on average. The methods of profound learning to solve the most difficult tasks in the fields of genetics, bioinformatics, biomedicine, robotics, and 3D technology work are clearly described in [23]–[26].

In [28], it is shown that one layer can be learned at a time from a densely connected deep assumption network. The easiest approach to achieve so is to presume that higher layers do not occur while studying lower layers, but that is not consistent with the usage of clear factorial approximations to substitute an intractable posterior distribution. It is proposed that a multi-level DL approach cover land and crop types using multitemporal satellite imagery [29]. This architecture employs both unregulated and directed NNs for satellite imagery segmentation and succeeding classification correspondingly.

The research paper [30] was proposed for a system which classifies the healthy and disease crop, it captures the image of fruits, leave or stem and send it to analyze engine, and analysis is done by using CNN (Conventional Neural Network), it analyses the image and classify crops according to condition and analyzed result is sent to the farmer who requires the decision. And a diagnosis of the diseases is done with a sufficient set of image data using deep learning. It performs the accuracy for classification, therefore it increases productivity. Authors in [31] established three separate deep learning models to segment plant regions, count plants, and estimate biomass from photos from aerial fields. Results show better accuracy of biomass estimation than previous approachesand better accuracy for counting outdoor emergence compared to previous indoor leaf counting research.

Sharda P et al. projected a method for plant leaf disease detection with advanced machine learning techniques [27]. This study shows how a machine learning approach can be built to enable the automatic diagnosis of illness through the recognition of images. Using a public set of 54,306 images of diseased and healthy plant leaves, a large, convolutional neural network was trained to identify crop and disease species in 38 separate classes of 14 crops and 26 diseases, with an accuracy of over 95 percent. They primarily began using the Plant Village dataset in the new program. The study is conducted on 54,306 photos of plant leaves assigned to them, which have a distribution of 38 class marks. If the class label is a pair of crop diseases, and they make an attempt to predict the pair of crop diseases given only the plant leaf picture. They change the image size to 256×256 pixels, and on these downscaled images we execute both model optimization and model predictions. They used three types of dataset models. The first they began with the dataset for color images. For Plant Village, they used a gray-scaled version of the dataset. The final version of the dataset they used was segmented. Extra context image information could lead to some implicit bias in the dataset.

In Figure 3.3, we can find the leaf images after the segmentation process which are (a) Color version of Leaf 1, (b) Grayscale version of Leaf 1, (c) Segmented form of Leaf 1, (d) Color version of Leaf 2, (e) Grayscale version of Leaf 2, (f) Segmented form of Leaf 2. The above images show us the view of various stages of the segmentation process. The color image is converted into a grayscale image. Later segmented image is taken from the grayscale image. The three versions (color, grayscale, segmented) of the data set indicate a standard output fluctuation over all experimentsif the rest of the experimental setup is maintained continuously. For a colored version of the dataset, the models perform best. We required the neural networks when designing the experiments, to learn only to understand the inherent biases in lighting conditions, data collection system, and apparatus. Therefore, we have tried to examine the adaptability in the absence of color details in the grayscaled version of the same dataset, and the ability of the model to acquire

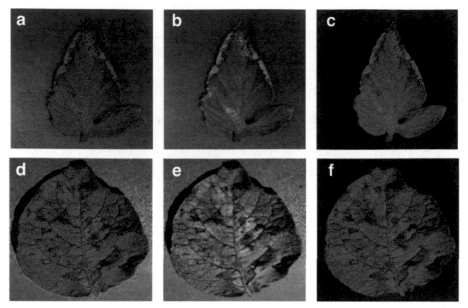

Figure 3.3   Various pair of crop disease pictures.

higher structural patterns for different crops and conditions. The results, as expected, decreased compared with the color version experiments of the dataset but the mean score observed was 0.852 even for the worst results, (85.2 percent overall accuracy).

As illustrated in Figure 3.3, The output of the segmented image model was always better than that of the gray-scaled image model, but slightly worse than that of the colored image model. The segmented versions in the entire dataset were prepared to analyze the position of images context in total results.

From the literature, it has been understanding that Plant diseases arise due to the intervention of infection. Pests are of many kinds thus many diseases that make crops unhealthy. Many pesticides are available which can prevent or destroy the invasion of infection. In this chapter, many crop diseases are identified and classified like Apple plant disease, Grape, Cherry, Peach, Corn, Pepper, Tomato, Strawberry, Potato plant disease, and so on. Table 3.2 depicts images of the leaves collected from the literature out of which some are considered for the simulation process in this proposal. The leaf class and the disease it attacked by are mentioned.

Table 3.2 Details of diseases occurs in different plant leaves

Leaf Image with Infection	Affects
 Apple leaf scab disease	The reason for apple scab disease is due to the existence of a fungus named Venturia inequalis. This disease is commonly found in the winter season because in the spring season rain disturbs this fungus and causes scabs. Although with the scab it's not harmful so it's still suggestable to eat apples because they are perfectly good on the inside of the apple and good to eat. It spreads rapidly as it is carried by wind, rain, or splashing water from the ground to flowers or fruit.
 Apple leaf black rot	The reason for the occurance of apple block rot is by a fungus named Botryosphaeria obtusa. It is also a serious disease of cultivated and wild grapes. This disease is very threatening in the warm and wet seasons. Thus, the reason for the occurance of black rot is moisture which can be overcome by providing great air circulation. It occurs early in the spring season when the leaves are unfolding. They appear as small, purple specks within 3 to 6 mm diameter on the upper parts of the leaf. Heavily infected leaves become necrotic and defoliation occurs.
 Apple leaf rust disease	It is caused by a fungal pathogen named Gymnosporangium juniperi- Virginians. This fungus attacks apples mostly in the season of spring. It looks like large yellow sports appear on cedar rust at the upper parts of the leaf turning yellow, orange to gray-green. Basidiospores, teliospores, spermatia, and aeciospores are the four types of spores produced by this rust. Teliospores are produced on gelatinous telial horns that develop from golf ball-like growths on red cedars and other junipers.
 Cherry leaf powder mildew disease	It grows up rapidly in environments of high humidity and moderate temperature. The main reason for powdery mildew is induced by fungal called podosphaera clandestine. It does not affect humans i.e. won't hurt if you touch it. It cannot sustain at high temperatures. The strong sun rays kill spores before they can spread.

(continued)

Table 3.2   (*Continued*)

 Grape leaf isariopisis disease	Isariopisis is fungal from the family of Mycosphaerellaceae which is called an isariopsis leaf spot caused by Pseudocercospora.
 Corn Leaf Spot disease	It is viewed as a foliar sickness and can be particularly destroying the sugar beet crops. Sometimes this infection is misdiagnosed as a dark spot. The disease will start at the base of the plant and will develop toward leaves with new development. This happens when the contagious spores develop and enter through characteristic openings of leaves in the ideal conditions. Symptoms will appear only after 5 to 12 days after spores infect a plant
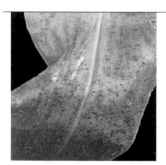 Corn (maize)leaf common rust	This fungus mainly survives in the winter season. It's growing rapidly in moist and cool conditions. Whereas in hot and dry conditions it is typically slow or stops growing. Although a few rust pustules can generally be found in corn fields throughout the growing season while syndrome generally does not appear before tasseling. These can be recognized effectively from other diseases by the development of dark, reddish-brown pustules found on both the upper and lower surfaces of the corn leaves.
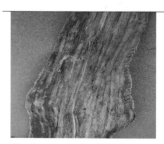 Northern corn leaf blight	It is due to the fungus Exserohilum turcicum and is one of the most often occurring foliar infections of corn. And it's symptoms usually occur first on the lower leaves. This can develop on leaves that differ from one to seven inches in length. These sores will initially show up as narrowtan marks that run corresponding with the leaf veins. As the sickness advance, the injuries become together in bigger regions of dead leaf tissue.

Table 3.2 *(Continued)*

 Grape leaf black rot	The main reason for this disease is created by an ascomycetous fungus, Guignardia bidwellii that assaults grapevines in hot and humid weather. And this infection also affects other parts of the plant. In order to reduce the black rot is by increasing the amount of moisture. A few plants give almost no indications of contamination until it's past the point of no return. They will look extremely sound until natural product sets. Indeed, even the blossoming will be typical. Contamination of the natural product is the most vital period of the sickness and will bring about financial misfortune. a greater amount of the plant.
 Grape leaf esca black measles	It is nothing but a black spot found on the surface of grape leaves. It can occur any time from the fruit set to days of harvest. Infected fruits will dry and it gives a very bitter taste called esca. It frequently occurs in seasons of heavy rains and high temperatures in the growing season. Leaf symptoms are recognized by a tiger stripe pattern while infections grow severe from year to year.
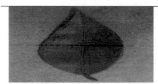 Orange leaf Citrus disease	It is one of the destructive diseases of citrus worldwide. Citrus huanglongbing (HLB) is formerly referred to as a citrus greening disease. In this citrus disease influences the tree health thereby the growth of fruit, ripening and worth of citrus fruits and juice.
 Peach leaf bacterial spot disease	The bacterial spot of Peach infects the fruit by devitalization of trees and these trees are further affected by winter injury. It is also a formation of cankers. Side effects of this illness incorporate natural product spots, leaf spots, and twig infections. During the rainy season it's very difficult to control the disease spread as the bacteria spreads vigorously.

(continued)

Table 3.2   (*Continued*)

 Tomato mosaic virus	In young plants, the infection reduces gradually. Sometimes an entire plant may be dwarfed and the flowers are discolored. In this environmental condition influence the symptoms.
 Pepper leaf bell bacterial spot	In Bell pepper spots small circular pimples are formed on the lower and upper surface of the leaves. They have similar symptoms on the tomato crop also. mild infection of bacteria will lead to easily seen necrotic spots on leaf's and a serious infection will cause immature leaf drops and pods that result in unsellable fruit. Like most bacterial diseases, the bacterial spot is difficult to control in the rain and moist conditions.
 Potato leaf early blight disease	Early blight is one of the potato diseases because of the fungus named Alternaria solani. In this disease, small circular dark brown or black spots are formed on the lower part of the leaf and later they will spread. The disease initially affects leaves and stems but only in supporting weather conditions and if left without taking precautions can result in considerable defoliation and enhancement of infection.
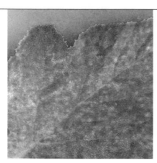 Squash leaf powdery mildew disease	During summer this disease is mostly prone to attack the plant specifically on the butternut and spaghetti squash. For this moist condition are not necessary at all to foster this fungus. This appears on older leaves first as reddish-brown spots but in the early stages, it can only be identified via microscope. This makes the leaves appear to have been dipped in talc. The leaves lose their normal dark green and turn pale yellow later to brown and finally shrivel. This disease can spread through the air and it spreads from plant to plant reducing the crop yield.

Table 3.2   (*Continued*)

Strawberry leaf scorch

Singed strawberry leaves are brought about by a parasitic disease that influences the foliage of strawberry plantings. The parasite mindful is called diplocarpon. Strawberries with leaf singe may initially give indications of an issue with the improvement of little purplish flaws that happen on the topside of leaves.

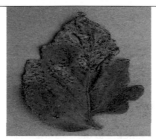

Tomato-leaf bacterial-spot disease

Bacterial spot is caused by four species of Xanthomonas and transpires worldwide wher/ever tomatoes are cultivated. These are small brown spots with yellow around the spot. Bacterial spot causes leaf and fruit spots which leads to the loss of yield. Due to diversity within the bacterial spot pathogens, it can occur at different temperatures which makes it a hazard to tomato production worldwide. This is similar to bacterial speck and impossible to identify in the initial stages.

Tomato septoria leaf spot

A very common disease of tomatoes is Septoria leaf spot. It generally appears in areas with high humid content. This causes an immense amount of destruction to the yield and considerable losses are faced on account of this
disease. It spreads quickly to defoliate and debilitate the plants making them unfit to endure natural product to development. This spreads from older leaves to young leaf's and the leaf turns yellow to brown and then wither.

Tomato leaf spider mites

Spider mites can ruin a tomato garden when they conquer and suck plant juices and then generally sit on the base of the leaf. These are one of the major infection problems of tomato plants. Grown primarily as grows in subtropical climates. These mites can be removed by opening a water hose on the plants with pressure enough to drive away from the mites off the leaf.

(continued)

Table 3.2   (*Continued*)

Tomato target spot

The disease starts on the grownup leaves and spreads upwards towards younger leaves that are just blooming. The initial signs are random-shaped spots that are inferior 1 mm and a yellow margin around them. While few spots may grow up to 10 mm and show attribute rings thus the name "target spot" is given. They also spread to other leaves rapidly causing the leaves to turn yellow and collapse. Spots will likewise happen on the stems that are long and thin.

Tomato yellow leaf curl virus

Tomato yellow leaf twist contamination is caused by the Begomovirus type. It is the most devastating tomato disease, and it is very likely to be found in warm areas, causing serious financial losses. This infection is spread by a creepy crawly vector belonging to the Aleyrodidae family. The indispensable host for TYLCV is a tomato plant and which can likewise be found in eggplants, potatoes, beans, and peppers. Specialists discovered contamination being obviously transmitted from infected guys to untainted females or bad habit.

Tomato healthy leaf

A tomato is a supplements thick superfood that offers an advantage to the limits of our body. Its nourishment's backings restorative skin, weight reduction, and heart wellbeing.

The data set contains various plant species such as apple, corn, grapes, potato, and tomato as shown in Table 3.2. The size of a dataset can be improved by creating converted types of images by applying horizontal flipping, vertical flipping, zooming, shearing, shift in width, shift in height. This happens using data augmentation methods in machine learning models. Training machine learning neural network models on more data will result in more skillful classification of models and augmentation techniques can create a variety of images that can enhance the ability of models to add what they have learned to new images.

The health of the plant itself can be represented by the leaves. Plants suffering from diseases normally have yellowish or brownish leaves, as well as spots on rotten areas. This data is what we'd like to extract from image data. We want features that are discriminative in feature space, have low dimensionsand are resilient to data variability when extracting features from data. Each leaf may affect different types of diseases. This model takes leaf image as an input. The user takes a picture of the affected leaf and gives the image as input to the model. In this model Given image undergoes preprocessing techniques like segmentation. Features are extracted in the form of a single column vector from the preprocessed image. The image is transferred through different hidden layers like convolution layer, ReLu layer, pooling layer, fully connected layer. Bypassing the image into the sequence of these layers will results in accurate results. Feature value vector is given to the CNN model which results in a matrix based on the feature values, the class with maximum value can be labeled for the image.

3.3 Present Work of Image Recognition Using Machine Learning for an Application of Leaf Disease Detection.jpg

This chapter presents the algorithm used for detecting the diseases of leaves and also depicts the various diseases of different leaves of plants using supervised learning using a training dataset of 87,867 images, a validation dataset of 17,527 images, and few test images for predicting the output. All the images used are of 256×256 and of .jpg format.

In this proposal, a model is created using Keras with the backend as Tensorflow. The Algorithm comes into play when we start training an artificial neural network. To train this network, basically need to solve an optimization problem, here the weights are optimized within the model which is a most important step in Neural networks to get the best predicted output. Each neuron in this neural network has an arbitrary weight assigned to it, and these weights are continuously updated during training to achieve their optimal values. These weights are optimized depending upon the optimization algorithm that is Stochastic gradient descent. When setting weights, the SGD's main goal is to minimize the given loss function, which means assigning weights in such a way that the loss function is as close to zero as possible. The loss will be minimized during training, once the model is given data and labels. In the present algorithm, the model is trained for identifying diseases in leaves by giving a set of images of diseased leaves along with the labels. These training images are to state whether these leaf images are infected or healthy.

When one image of an infected leaf is fed to the model then this image will pass through the entire network and the model will produce output by going through various functions like Convolution (conv2d), Max pooling, Relu at the last stage of each layer. Relu is an activation function used at a node to introduce nonlinearity. Rearmost layer in the architecture is Softmax to yield the outcome. This output is what algorithm thinks about the image, it actually consists of probabilities of an infected leaf or healthy leaf, it will assign a probability for diseased and healthy images at the output say 80% percent being diseased and 20% percent being healthy. This means the algorithm assigns a higher likelihood for the image being diseased rather than a healthy leaf. In this case, it is going to find the error between the predicted image of the algorithm and the true image. To make our model as accurate as possible to get good predictions, it is necessary to train the model by passing more data for many iterations. At this stage, the algorithm makes the model to learn from the given data.

Dataset of infected and disinfected images of the same size and format is created. All Images are converted into arrays using NumPy for making the machine understand the image data for displaying the characteristics of data. Based on the requirement, data manipulation is done using the Pandas library. Now all images are fed as input to a model which will go through the layers of the network. Now, this model is trained by using Scikit learn algorithm so that model is ready for identification of various diseases in the leaf. In this process, model is trained with each and every image several times which are passed through many secret layers of the neural network. When all the data is processed and training is completed, a file is saved with an extension .h5 which contains the weights of the trained model. That will help to load the weights to identify and classify the input data. In this step, the model is tested with test data that yields the output identifying the type of disease with the label of the leaf. The algorithm displays a report of stating precision, recall, f-1 score, and support values for a given input leaf image.

As shown in Figure 3.4, every neural network consists of many neurons and every neuron has its respective bias and weights. it is activated when that bias is true and connected to other neurons through many neurons activation communication.

This process is carried in the neural network till the output layer produces the output. The first step is to collect the database that has images of different leaves with and without diseases. Then train the model with the help of train

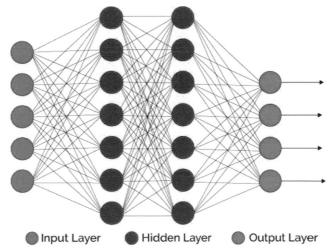

Figure 3.4 Neural Network.

and test dataset. To validate the model a new image of a leaf will be taken, that image will undergo segmentation to highlight the edges and boundaries. The model will analyze the extracted features and predicts whether the given image is normal or affected, if it is affected then it will predict the disease name. Once the model prediction and validation is completed, then the model gets ready for deployment as shown in Figure 3.5.

The same model can be used for the classification of different types of image data in addition to classifying diseases in plant leaves. In order to obtain the results, we use a test dataset. Thus, various diseases which are discussed above are identified and classified by our trained module. Some of the predicted images of the proposed model are shown in Figure 3.6.

Classification Report

Classification is a very imperative process here the leaves are classified based on the crop and then again classified based on their diseases.
Classification report has factors as follows

Precision
Recall
f1-score
support

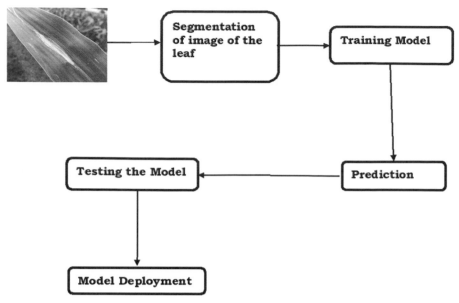

Figure 3.5 Flowchart of proposed method.

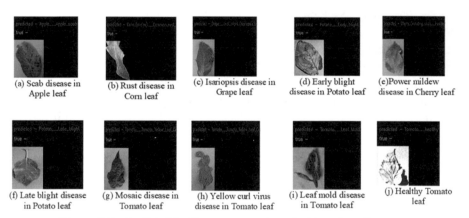

(a) Scab disease in Apple leaf

(b) Rust disease in Corn leaf

(c) Isariopsis disease in Grape leaf

(d) Early blight disease in Potato leaf

(e)Power mildew disease in Cherry leaf

(f) Late blight disease in Potato leaf

(g) Mosaic disease in Tomato leaf

(h) Yellow curl virus disease in Tomato leaf

(i) Leaf mold disease in Tomato leaf

(j) Healthy Tomato leaf

Figure 3.6 Predicted leaf images contaminated with various diseases.

The maximum value of these factors is one. In the output, we get the name of the object that has been trained followed by precision.

Precision: Precision is the accuracy of the leaf that has been detected in our case that is a leaf. Value it shows is the value of how sure the computer is about the output however; this value can be increased by increasing the training data. Precision is given by the formula:

Precision = True Positives / (True Positives + False Positives)

Recall: Recall is the value of positive and negative instances. Positive and negative are the what computer thinks about the input or test images. Recall is given by the formula:

Recall = True Positives / (True Positives + False Negatives)

F-1 Score: It is the average score of each class. It is given by the following formula:

F-1 Score = 2 (Precision * Recall) / (Precision + Recall)

Support: It is the valid data set that we give for a system to learn. The data sets we have given to train and valid data are compared during the learning process. A classification report is a detailed report that is used to evaluate a model's prediction.

A classification report is a detailed report that is used to assess a model's prediction. It is one of the most basic steps in machine learning. After training a machine learning model it is major to calculate the model's achievement. One approach to do this is by using SKlearn's classification report. This classification report table is built from a confusion matrix when input is given to the model it gives an output then these are tabulated in the table for each input data. Then these values are classified into 4 types True positives, True negatives, False negatives, False positives as shown in Table 3.2 for performance measurement. Accuracy is calculated by the sum of True positives divided by the total dataset.

Accuracy will increase only when the True positives number are high. F-1 score is a metric for imbalanced data. It considers both rows(recall) and columns (precision). F-1 score is the harmonic mean of the recall and precision. When precision is very high compared to recall or vice versa then it is called imbalanced data. Then Harmonic data punishes the extreme data by which harmonic average is calculated and the f-1 score is tabulated.

Precision is calculated by identifying the True positives and False positives and assigning these values to a precision formula. The average precision is calculated by True positives divided by the sum of True positives and false positives by total classes. A recall is calculated by True positives and False negatives and their average is calculated by True positives divided by the sum of True positives and False negatives by total. Precision, recall, f-1 score, and support are measured using their respective formulas. Table 3.3 demonstrates the classification report of specific leaves that are used in this model stating the values of precision, recall, f1-score, and support.

Table 3.3   Classification report of the proposed model

	Precision	Recall	fl-Score	Support
Apple scab	0.78	0.34	0.47	504
Apple Black rot	0.45	0.86	0.59	497
Apple Cedar apple rust	0.95	0.30	0.46	440
Apple healthy	0.63	0.49	0.55	502
Blueberry healthy	0.75	0.27	0.40	454
Cherry (including sour) Powdery mildew	0.94	0.26	0.41	421
Cherry (including sour) healthy	0.91	0.45	0.60	456
(maize)_Cercospora leaf spot Gray leaf_spot	0.59	0.36	0.45	410
Corn (maize) Common rust	0.98	0.80	0.88	477
Corn (maize) Northern Leaf Blight	0.84	0.58	0.69	477
Corn_(maize)_healthy	1.00	0.70	0.82	465
Grape Black rot	0.69	0.28	0.39	472
Grape Esca (Black Measles)	0.60	0.66	0.63	480
Grape Leaf blight (Isariopsis Leaf Spot)	0.69	0.40	0.50	430
Grape healthy	0.91	0.07	0.14	423
Orange Haunglongbing (Citrus greening)	0.95	0.60	0.74	503
Peach Bacterial spot	0.83	0.42	0.56	459
Peach healthy	0.70	0.86	0.77	432
Pepper bell Bacterial spot	0.43	0.20	0.27	478
Pepper bell healthy	0.59	0.26	0.36	497
Potato Early blight	0.39	0.74	0.51	485
Potato Late blight	0.39	0.27	0.32	485
Potato healthy	0.31	0.49	0.38	456
Raspberry healthy	0.84	0.18	0.30	445
Soybean healthy	0.68	0.64	0.66	505
Squash Powdery mildew	0.67	0.54	0.60	434
Strawberry Leaf scorch	0.56	0.52	0.54	444
Strawberry healthy	0.69	0.64	0.66	456
Tomato Bacterial spot	0.76	0.09	0.16	425
Tomato Early blight	0.33	0.46	0.39	480
Tomato Late blight	0.48	0.41	0.44	463
Tomato Leaf Mold	0.30	0.56	0.39	470
Tomato Septoria leaf spot	0.21	0.37	0.27	436
Tomato Two-spotted spider mite	0.24	0.40	0.30	435
Tomato Target Spot	0.19	0.33	0.24	457
Tomato Yellow Leaf Curl Virus	0.69	0.54	0.61	490
Tomato mosaic virus	0.27	0.71	0.39	443
Tomato healthy	0.23	0.97	0.38	481

3.4 Conclusion

Harvest security in natural agribusiness is anything but a basic issue. It relies upon a careful information on the yields developed and their probable vermin, pathogens and weeds. Currently, machine learning approaches are extremely vigorous to practical environments and the structures certainly have the benefit of the learning process. In our task, a particular profound learning model was formed dependent on obvious convolutional NN structures for the discovery of leaf infections through leaves depictions of infected or healthy leaves. The preprocessed leaf image information is passed to machine learning models. The training and validation datasets are used to train and tune the machine learning models. Once the parameters of the model are optimized, the model is chosen for image classification to detect whether the leaf is infected or not. Our test results and examinations between different profound models with include extractors showed how our profound learning-based indicator can effectively perceive various classifications of illnesses in different leaf's and furthermore give an answer for concerned diseases. The proposed Scikit-learn strategy is applied to characterize and distinguish the security of the pictures and to expand the creation of farming by recognizing the disease in the harvest. This task has proposed a constant identification approach that depends on improved convolutional neural systems for different leaf sicknesses. The profound learning-based methodology can naturally separate the discriminative highlights of the infected harvests like apple, corn, grape, potato, peach, pepper, potato, strawberry, and potato leaf pictures and identify their different leaf sicknesses with high precision progressively. This venture is to guarantee agreeable and speculation execution of the proposed model and an adequate infection picture dataset aggregate of 87,867 pictures with uniform and complex foundations were gathered. Besides, the new profound convolution neural system model utilizing SK learn as planned an Inception module and coordinating with Keras and TensorFlow to upgrade the multi-scale ailment object identification and little infected item location exhibitions. Utilizing an approval dataset of 17,527 pictures of infected leaves the proposed model was prepared to identify leaf ailments. The thorough discovery execution arrives at 0.7. Henceforth, the prepared model is completely fit for Real time identification of different prepared leaf illnesses. The outcomes exhibit that the proposed AI model can distinguish the expressed kinds of leaf illnesses with high exactness continuously and gives a doable answer for the constant location of leaf sicknesses. In the future, we trust our working model will make an interesting commitment to farming by designing a cost-effective bot. The moisture in the soil is also identified by giving inputs to bot for spraying water and pesticides. The moisture in the soil was also identified by using 222- HMP60A because this can be easily replaced by the farmer himself and its membrane filter protects itself from dirt & dust.

References

[1] Jiang Lu, Jie Hu, Guannan Zhao, Fenghua Mei, Changshui Zhang, "An in-field automatic wheat disease diagnosis system", Computers and Electronics in Agriculture, Vol. 142, Pages 369–379, 2017.

[2] Andreas Kamilaris, Francesc X. Prenafeta-Boldu, "Deep learning in agriculture: A survey", Computers and Electronics in Agriculture, Vol. 147, Pages 70–90, 2018.

[3] Konstantinos P. Ferentinos, "Deep learning models for plant disease detection and diagnosis", Computers and Electronics in Agriculture, Vol. 145 Pages 311–318, 2018.

[4] Kulkarni Anand H, Ashwin Patil RK, "Applying image processing technique to detect plant diseases", Int J Mod Eng Res, Vol. 2, Issue 5, Pages 3661–3664, 2012.

[5] Bashir Sabah, Sharma Navdeep, "Remote area plant disease detection using image processing", IOSR J Electron Commun Eng ISSN, Vol 2 Issue 6, Pages 2278-2834, 2012.

[6] Rakesh Kaundal, Amar S Kapoor and Gajendra PS Raghava "Machine learning technique in disease forecasting: a case study on rice blast prediction," BMC Bioinformatics, 2006.

[7] S. Sankaran, A. Mishra, R. Ehsani, and C. Davis, "A review of advanced techniques for detecting plant diseases," Computers and Electronics in Agriculture, vol. 72, Issue 1, Pages1–13, 2010.

[8] P. R. Reddy, S. N. Divya, and R. Vijayalakshmi, "Plant disease detection techniquetool-a theoretical approach," International Journal of Innovative Technology and Research, Pages 91–93, 2015.

[9] A.-K. Mahlein, T. Rumpf, P. Welke et al., "Development of spectral indices for detecting and identifying plant diseases," Remote Sensing of Environment, vol. 128, Pages 21–30, 2013.

[10] W. Xiuqing, W. Haiyan, and Y. Shifeng, "Plant disease detection based on near-field acoustic holography," Transactions of the Chinese Society for Agricultural Machinery, vol. 2, article 43, 2014.

[11] A.-K. Mahlein, E.-C. Oerke, U. Steiner, and H.-W. Dehne, "Recent advances in sensing plant diseases for precision crop protection,"

European Journal of Plant Pathology, vol. 133, Issue 1, Pages 197–209, 2012.

[12] P. Chaudhary, A. K. Chaudhari, A. N. Cheeran, and S. Godara, "Color transform based approach for disease spot detection on plant leaf," International Journal of Computer Science and Telecommunications, vol. 3, Issue 6, Pages 65–69, 2012.

[13] S. B. Patil and S. K. Bodhe, "Leaf disease severity measurement using image processing," International Journal of Engineering and Technology, vol. 3, Issue 5, Pages 297–301, 2011.

[14] J. K. Patil and R. Kumar, "Feature extraction of diseased leaf images," Journal of Signal & Image Processing, vol. 3, Issue 1, Pages 60, 2012.

[15] T. R. Reed and J. M. H. Dubuf, "A review of recent texture segmentation and feature extraction techniques," CVGIP: Image Understanding, vol. 57, Issue 3, Pages 359–372, 1993.

[16] M. S. P. Babu and B. Srinivasa Rao, "Leaves recognition using back propagation neural network-advice for pest and disease control on crops," IndiaKisan, Expert Advisory System, 2007.

[17] P. Revathi and M. Hemalatha, "Identification of cotton diseases based on cross information gain_deep forward neural network classifier with PSO feature selection," International Journal of Engineering and Technology, vol. 5, Issue 6, Pages 4637–4642, 2014.

[18] C. Zhou, H. B. Gao, L. Gao, and W. G. Zhang, "Particle swarm optimization (PSO) algorithm," Application Research of Computers, vol. 12, Pages 7–11, 2003.

[19] T. Rumpf, A.-K. Mahlein, U. Steiner, E.-C. Oerke, H.-W. Dehne, and L. Plümer, "Early detection and classification of plant diseases with Support Vector Machines based on hyperspectral reflectance," Computers and Electronics in Agriculture, vol. 74, Issue 1, Pages 91–99, 2010.

[20] Z. H. Zhou and S. F. Chen, "Neural network ensemble," Chinese Journal of Computers, vol. 25, Issue 1, Pages 1–8, 2002.

[21] B. C. Karmokar, M. S. Ullah, Md. K. Siddiquee, and K. Md. R. Alam, "Tea leaf diseases recognition using neural network ensemble," International Journal of Computer Applications, vol. 114, Issue 17, Pages 27–30, 2015.

[22] H. Al-Hiary, S. Bani-Ahmad, M. Reyalat, M. Braik, and Z. AL-Rahamneh, "Fast and accurate detection and classification of plant diseases," Machine Learning, vol. 14, Pages 5, 2011.

[23] I. Lenz, H. Lee, and A. Saxena, "Deep learning for detecting robotic grasps," The International Journal of Robotics Research, vol. 34, Issue 4-5, Pages 705–724, 2015.

[24] B. Alipanahi, A. Delong, M. T. Weirauch, and B. J. Frey, "Predicting the sequence specificities of DNA- and RNA-binding proteins by deep learning," Nature Biotechnology, vol. 33, Issue 8, Pages 831–838, 2015.

[25] L. Zhang, G.-S. Xia, T. Wu, L. Lin, and X. C. Tai, "Deep learning for remote sensing image understanding," Journal of Sensors, vol. 2016, Article ID 7954154, Pages 365-366, 2016.

[26] J. Arevalo, F. A. Gonzalez, R. Ramos-Pollan, J. L. Oliveira, and M. A. G. Lopez, "Convolutional neural networks for mammography mass lesion classification," in Proceedings of the 37th Annual International Conference of the IEEE Engineering in Medicine and Biology Society (EMBC '15), Pages 797–800, August 2015.

[27] Sharda P. Mohanty, David P. Hughes, Marcel Salathe, "Using Deep Learning for Image-Based Plant Disease Detection" Frontiersin Plant Sci, Vol. 7, Issue 4 2016

[28] Asa Ben-Hur, Jason Weston "A Robust Deep-Learning-Based Detector for Real-Time Tomato Plant Diseases and Pests Recognition", Sensors, Vol. 4, Issue 6 2017

[29] Nataliia Kussul, Mykola Lavreniuk, Sergii Skakun, and Andrii Shelestov "Deep Learning Classification of Land Cover and Crop Types Using Remote Sensing Data", IEEE, Vol. 14, Issue 5 2017

[30] Hyeon Park, Eun JeeSook and Se-Han Kim "Crops Disease Diagnosing using Image-based Deep Learning Mechanism", International Conference on Computing and Network Communications (CoCoNet), IEEE, Aug 2018

[31] Shubhra Aich, Anique Josuttes, Ilya Ovsyannikov, Keegan Strueby, Imran Ahmed, Hema Sudhakar Duddu "DeepWheat: Estimating Phenotypic Traits from Crop Images with Deep Learning", Winter Conference on Applications of Computer, IEEE, March 2018

4

COVID-19 Forecasting Using Deep Learning Models

**Sahiti Cheguru[1], Prerana CH[2], K. Tejasree[3], Tarine Deepthi[4],
Y. Vijayalata[5], and Ghosh Siddhartha[6]**

[1]Gokaraju Rangaraju Institute of Engineering and Technology, India
[2,3,4,6]Vidya Jyothi Institute of Technology, India
[5]Gokaraju Rangaraju Institute of Engineering and Technology, India
E-mail: sahiticheguru2000@gmail.com; preranacheguru@gmail.com;
tejasreekomati05@gmail.com; tarinedeepthi1998@gmail.com;
vijaya@griet.ac.in; siddhartha@vjit.ac.in

Abstract

COVID-19, responsible for infecting billions of people and the economy across the globe, requires a detailed study of the trend it follows to develop adequate short-term prediction models for forecasting the number of future cases. In this perspective, it is possible to develop strategic planning in the public health system to avoid deaths as well as managing patients. In this paper, forecast models comprising various artificial intelligence approaches such as support vector regression (SVR), long short term memory (LSTM), bidirectional long short term memory (Bi-LSTM) are assessed for time series prediction of confirmed cases, deaths, and recoveries in ten major countries affected due to COVID-19. The paper also reviewed a deep learning model to forecast the range of increase in COVID-19 infected cases in future days to present a novel method to compute multidimensional representations of multivariate time series and multivariate spatial time series data. The paper enables the researchers to consider a large number of heterogeneous features, such as census data, intra-county mobility, inter-county mobility, social distancing data, past growth of the infection, among others, and learn complex

77

interactions between these features. To fast-track further development and experimentation, the analyzed code could be used to implement the AI in an efficient way. The paper discusses existing theories and research that provide a better understanding of the spread pattern recognition which will help to tackle any future pandemic of similar intensity. We encourage others to further develop a novel modeling paradigm for infectious disease based on GNNs and high resolution mobility data.

Keywords: COVID-19, Artificial intelligence, Deep learning model, Time series data, Prediction model

4.1 Introduction

Corona viruses earn their name from the characteristic crown-like viral particles (virions) that dot their surface. This family of viruses infects a wide range of vertebrates, most notably mammals and birds, and are considered to be a major cause of viral respiratory infections worldwide [1]. With the recent detection of the 2019 novel corona virus (COVID-19), there are now a total of 7 corona viruses known to infect humans. Prior to the global outbreak of SARS-CoV in 2003, HCoV-229E and HCoV-OC43 were the only corona viruses known to infect humans. Following the SARS outbreak, 5 additional corona viruses have been discovered in humans, most recently the novel corona virus COVID-19, believed to have originated in Wuhan, Hubei Province, China. COVID-19 effect has highly noticeable in dense areas with elderly people and people with co-morbidities [2]. It is considered a multidisciplinary issue for the medical specialists, pharmaceutical industry, local government/health authorities, and epidemiological experts. This study is mainly focused on the review of forecasting and prediction of COVID-19 using various deep learning algorithms. A big challenge has been witnessed in various science domains globally to restrict the increasing COVID spread trends. Various modeling, forecasting, and analysis approaches are established to handle and insight this current pandemic. The evolution of confirmed COVID cases forecasting has been estimated by multiple mathematical models [3, 4].

This study is mainly focused on the review of forecasting and prediction of COVID-19 using various deep learning algorithms. A big challenge has been witnessed in various science domains globally to restrict the increasing COVID spread trends. Various modeling, forecasting, and analysis approaches are established to handle and insight this current pandemic.

The evolution of confirmed COVID cases forecasting has been estimated by multiple mathematical models. This study is aimed at deep learning models and a comparative study is made for forecasting COVID-19 cases. The deep learning models such as long short term memory LSTM, Bidirectional LSTM, Gated Recurrent unit- GRU, and Recurrent neural network- RNN have been analyzed. These models possess various advantages like distribution free learning models, managing temporal dependencies in time series data, and nonlinear features modeling of flexibility. Various datasets have been utilized in various studies like the John Hopkins dataset from starting to now COVID-19 status. The comparative study and challenges are exhibited in this study.

The major contribution of this study involves,

- To review the various deep learning models related to COVID-19 forecasting and time series prediction globally.
- To analyze the LSTM, Bi-LSTM, and GRU techniques applied in various medical images related to COVID-19 cases.
- To make a comparative study for the discussion related to COVID-19 prediction and forecasting.

The following Section 4.2 describes the deep learning models against covid-19 and their applications, Section 4.3 describes the population attributes of COVID-19, followed by Section 4.4 describes the various deep learning models and the involved COVID-19 dataset. Finally, the conclusion is presented in Section 4.5.

4.2 Deep Learning Against Covid-19

With the regular increase in the newly acquired and suspected COVID-19 cases, diagnosis of the disease is becoming a growing issue in most of the main hospitals because of the inadequate supply of detection systems in the corresponding epidemic area. Radiography and computed tomography hence originated as the integrative players in the pre-detection and diagnosis of COVID-19. But due to the aforementioned overwhelming patients, false positive rates leading to urgent requirements of computer automated diagnosis like deep learning that precisely confirm patientsscreen them thereby conducting viral surveillance. The following studies developed a deep learning process on the basis of CT diagnosis for the detection of COVID-19

patients that were able to automatically retrieve the radiographic characteristics of the novel virus, particularly the GGO (ground glass opacity) from
the radiographic images.

This research [5] developed a DL framework for the automatic quantification and segmentation of the quantification of the infectious areas and
the whole lung from the corresponding chest scans. The paper employed
VB-Net NN (neural network) for the segmentation of COVID-19 infection
areas in CT images. This setup has been trained with the utilization of two
hundred and forty-nine COVID patients followed by the validation of three
hundred patients. For accelerating the manual description of CT images to
train the features, a HITL (human in the loop) has been adapted for assisting
the physician in refining automatic annotation in every case. The assessment
of the DL based performance system in accordance with Dice similarity
coefficient, percentage of infection in between the manual and automatic
segmentation outcomes on the validated images.

Moreover [6] the study provided a fully automated and rapid diagnosis
of COVID-19 by adopting deep learning. The experimental assessment on
6524 X-rays of various institutions described the efficiency of the suggested
method with an average detection time of 2.5 seconds as well as with average
accuracy of 0.97 [7]. Formulated the task of classifying viral pneumonia
from the healthy controls and non-viral pneumonia into anomaly detection
problems. Hence the study suggested a CAD model that consisted of shared
feature extraction, prediction module, and detection module. The main benefit
of the suggested method over the binary classification is preventing individual
class explicitly followed by the complete treatment. This suggested model
possesses greater efficiency of AUC 84% and a sensitivity of 72%.

Similarly [8] evaluated the longitudinal modifications of pneumonia in
various COVID-19 clinical types at the baseline and follow-up imaging
with the use of quantitative image parameters that has been automatically
developed by deep learning systems from chest X-rays. The major findings of
the study are lung opacity burden, entire lung, and per lobe comparison. This
system was able to assess quantitatively the percentage of lung pacification
and the recent vision required for the radiologist supervision. The study
yielded 8.7% of the cases for insufficient segmentation that ensures precise
quantification.

4.2.1 Medical Image Processing

Medical image processing is a complex method and understanding of this
process is the main cause in the patients who do not respond to the CRT.

The study [9] demonstrated the voltage dependent right ventricle capture by the misplaced right atrial lead. The study suggested that device interrogations with the 12 lead ECG and succeeding multimodality imaging must be regarded in accordance with the premature diagnosis of non-responder.

[10] The study aimed to offer burnout medical professions an opportunity by intelligent DL classification methods. The study detected an appropriate CNN model by an initial comparative analysis of various CNN frameworks. The study then optimized the selected VGG 19 model for image modeling for depicting that the model might be utilized for high demand and challenging datasets. The paper then highlighted the limitations in using the publicly available datasets for the development of useful DL models and the process of creating an adverse impact on training the complex system. The study also suggested an image pre-processing stage for creating a trustworthy dataset in order to develop and test the DL models. This robust method has been aimed to decrease the unwanted noise from the images thereby DL models could focus on identifying diseases with peculiar features from the extraction. The results represented that the US images offer an extraordinary detection rate when compared with the CT and X-ray scans. These experimental outcomes signified that with the presence of limited data, many deep networks suffer for training effectively and provide low consistency when compared with the three used image models. The selected model has been then widely tuned with the corresponding parameters and made to perform the COVID-19 detection over pneumonia or normal lungs for all the three lung models with the accuracy of 84% of CT, 100% of US, and 86% of X-ray.

Advanced AI methods [11] and deep learning techniques have depicted high efficiency in the detection of patterns like diseased tissue. This study examined the efficiency of the VGG 16 base DL model for the detection of COVID-19 and pneumonia with the employment of torso radiographs. The results depicted that a high level of sensitivity in the detection of COVID-19 associated with the high level of specificity represented that this model could effectively be used as the screening test. ROC and AUC Curves are higher than 0.9 for all the considered classes.

4.2.2 Forecasting COVID-19 Series

This article [12] employed six machine learning methods such as CUBIST, RIDGE, RF, SVR, and stack ensemble learning and ARIMA model for the cumulative confirmation of COVID-19 in the Ten Brazil States in accordance with the incidence. The study evaluated the stability of the efficiency and out of sample errors by box plots. The study failed to adopt the DL approach in combination with ensemble learning. The study did not attempt couples function for dealing with data augmentation. Also, the study adapts hyper parameter tuning forecasting of the upcoming cases of COVID-19.

On the other hand [13] focused on two main problems which are as follows: One which generates real time forecast of the upcoming COVID-19 case for several countries and next is the assessment of the risk of novel COVID-19 for few more affected countries by a determination of several significant demographic features of the countries and its disease characteristics. To resolve the initial problem, the study presented a hybridized approach on the basis of an autoregressive integrated moving model and a wave-let based forecast model for generating short-term forecasts to determine future predictions of the outbreak. This study might be useful for the efficient allocation of medical professionals and also it acts as an early warning framework for the government policy makers. The next issue could be solved by the application of optimal regression of tree algorithms in order to determine the important causative variables which considerably affect the fatal rates for various countries. This analysis would necessarily 4offer deep insight for understanding the early risk of assessing 50 highly affected countries.

4.2.3 Deep Learning and IoT

Because of the global pandemic, there is an emergency requirement for the utilization of technology to its optimum potential. IoT is considered as one of the recent methods with great capability in performing against the COVID-19 outbreak. IoT comprises a limited network where IoT devices sense the surrounding environment and send useful data on the internet. This research [14] examined the present status of IoT applications in relation to novel viruses for the identification and deployment of their operational challenges and suggested the possible outcomes for further pandemic situations. Apart from that, the study performed statistical analysis for the implementation of IoT where the external and internal factors are being discussed.

Likewise [15] tested several number of COVID-19 diagnosis methods that depend on deep learning algorithms with the corresponding instances. The test results of the study depicted that DL models did not consider defensive frameworks against adverse probabilities that remain vulnerable to the corresponding attacks. At lasts the study presented in detail regarding the implementation of the attack model of the prevailing COVID-19 diagnostic applications. The study hoped that this process will generate awareness of

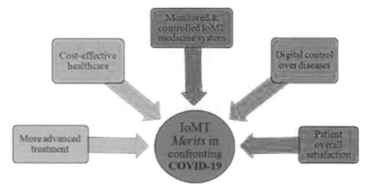

Figure 4.1   IoT merits towards COVID-19 [16].

the adversarial attacks thereby encouraging others to safeguard DL methods from the attack of the healthcare system.

In this article [16], the insight of DL tool application from the diverse view for empowering IoT applications in 4 major domains comprising smart home, smart health care, smart industry, and smart transportation is investigated. The main thrust has to be seamlessly coinciding with the two divisions of DL and IoT that resulted in an expensive range of new frameworks in the application of IoT like health monitoring, indoor localization, disease analysis, intelligent control, traffic monitoring, home robotics, autonomous driving, traffic prediction, and manufacture inspection. The study discussed the problems, future research, and challenges that use DL and for the motivation regarding further improvement in the promising area.

4.2.4 NLP and Deep Learning Tools

This study [17] utilized an automated extraction of the corona virus discussion from the social media and NLP method on the basis of topic modeling for uncovering several issues in accordance with the viral symptoms from public opinion. Further, the study also investigated the usage of LSTM RNN for the sentiment classification of COVID-19 comments. The findings of the present study focussed on the significance of the decision-making of COVID-19 issues.

Moreover [18] detected and analyzed sentiment emotions and polarity that has been described during the beginning of the initial stage of the pandemic lockdown period with employment of NLP and DL techniques on Twitter posts. LSTM models utilized for the estimation of emotions and sentiment polarity from the tweets extracted were trained to obtain existing accuracy on the sentiment 140 dataset. This use of emotions depicted a novel and unique method of estimating and validating the supervised learning models on the tweets extracted from Twitter.

4.2.5 Deep Learning in Computational Biology and Medicine

Advances in technology in imaging and genomics led to the explosion of cellular and molecular profiling of the data from huge numbers of samples. This tremendous rise of the biological data acquisition and dimension rate is a complex and conventional analytical strategy. The modern ML methods like deep learning promise to handle huge datasets for the determination of the hidden structure within them thereby making precise predictions. The review discussed the application of novel breeds and approaches in cellular imaging and regulatory genomics. The study provided a background of the summary of deep learning and provided certain tips for the practical usage with possible pitfalls and challenges for guiding the computational biologists in the utilization of this methodology [19]. The study [20] briefly introduced the following manuscripts and discussed their overall contribution to the advancement of science and technology: transcriptomic, cancer informatics, visualization, and tools, computational algorithms, micro biome research, and deep learning.

4.3 Population Attributes – Covid-19

This study emphasized the impact of COVID-19 for the migrant workers who are affected immensely. The geographical assessment analysis has been focused and the key facts to control this epidemic have been stated. The population attributes are shown in Figure 4.2. The structural barriers have been addressed. The intervention focal points are recognized by built environments and social networks. The risky role of migrant workers in Singapore is thus identified by the network's protective roles [28] . The public health and world economy were highly affected due to the COVID-19 pandemic. This kind of issue has been controlled by non-pharmaceutical interventions and this study utilized the Susceptible Exposed Infected Recovered-SEIR for pandemic dynamics simulation utilizing the society following government, people, and business. With respect to social cooperation, the higher realistic implementation related to various social interventions followed. Further COVID ABS models have been developed in the Python language.

Table 4.1   Comparative Study of the Prevailing Literatures

S.No	Author	Description and Methodology	Comments on the Results
1.	[21]	The study introduced a novel DL framework (COVIDX-NET) for assisting the radiologists in the automatic detection of corona virus presence in X-Ray images. This suggested framework comprised 7 various architectures of deep CNN like VGG 19 and the Google MobileNet (second version)	The study described the useful implementation of DL models for the classification of COVID-19 in the COVIDNet processed X-Ray images and supported further research in deep learning for diagnosing COVID-19 with high accuracy.
2.	[22]	The study utilized the DL model for the automated identification of anomalies in chest CT of COVID-19 patients and compared the quantitative estimation with the radiological residents. A deep learning algorithm consisting of detection of lesions, segmentation, and location has been trained and validated in 14,435 patients with definite pathogenic inclusion.	The suggested algorithm depicted excellent efficiency in the detection of COVID-19 pneumonia on the chest CT when compared with the existing radiologists.
3.	[23]	The issue of automatic classification of pulmonary diseases, comprising the recently emerged COVID-19, from X-ray images has been focussed in the study. In specific the existing CNN known as the Mobile net has been employed and trained from the scratch for the investigation of the significance of the features extracted for the classification task.	The results suggested that training CNN from scratch revealed vital biomarkers but not constrained to the COVID-19 disease, whereas the top classification accuracy suggested further analysis of the X-ray imaging potential.
4.	[24]	The paper assessed the usefulness of the (ARIMA) model in the prediction of the dynamics of Covid-19 incidence at various stages of the epidemic, from the intial growth phaseto the maximum daily incidence, until the phase of the epidemic's extinction	The study recommended the ARIMA model for forecasting COVID-19 for countermeasures.

Continued

Table 4.1   *Continued*

S.No	Author	Description and Methodology	Comments on the Results
5.	[25]	The study developed a prototype of a decentralized IoT based biometric face detection framework for cities under lockdown during the COVID-19 pandemic. The study built a deep learning framework of multi-task cascading for the detection of the face.	The study proved that it has an edge over cloud computing architecture.
6.	[26]	The study built an automated tool known as COVID-19 sign sym that could extract symptoms with their eight factors (severity, body location, condition, uncertainty, temporal expression, negation subject, and course) from the clinical text.	The information extracted is also mapped to the standardized clinical concept in the general OHDSI model. The evaluations of the notes followed by the medical sayings describe promising outcomes.
7.	[19]	Explored the possibility of Zakat and Qardh-Al-Hasan as a financial method to handle the adverse impact of Corona virus on poor and SMEs. It resolved by proposing an Artificial Intelligence and NLP based Islamic FinTech Model integrated with Qardh- and Al-Hasan Zakat	The study revealed that Islamic finance has immense potential to overcome any kind of pandemic like COVID-19
8.	[27]	The study signified the difference and similarity in extensively utilized models in deep learning studies, by discussing their basic structures, and reviewing diverse disadvantages and applications	The study anticipated the work can serve as a meaningful perspective for future development of the suggested algorithm in computational medicine.
9.	[28]	The paper investigated the networks of non-work related activities in migrant workers to intimate the improvement of lockdown exit techniques and upcoming pandemic preparedness	The study recommended social and geospatial distance followed by avoiding mass gathering and it also encouraged the welfare of migrant workers.

Continued

Table 4.1 *Continued*

S.No	Author	Description and Methodology	Comments on the Results
		It was conducted with 509 migrant workers over the nation, and it evaluated dormitory attributes, mental health status and social ties, physical and COVID-19-related variables, and mobility patterns with the use of grid-based network questionnaires.	
10.	[29]	The study assisted the policy makers in taking required decisions in order to stop the pandemic spread; precise forecasting of the propagation of the disease is the paramount significance. The suggested method initially groups the countries possessing the same socioeconomic and demographic details as well the health sector indicators with the use of k means algorithm	The method obtained high accuracy in forecasting the daily cumulative viral cases.
11.	[30]	The study might be used to differentiate several respiratory patterns and the suggested device could be readily employed for practical utilization.	The suggested deep learning possesses the vital potential to be extended to large-scale applications like sleeping scenarios, public places, and office environments.

By modifying the input parameters this developed model can be extended to other populations/societies. For health and government authorities, this model is very helpful [31]. In Israel, 271 localities have been assessed during the outbreak of 3 months in which 90 percent of the population is urban. Higher infection rates were seen in political minority groups. On the urban political attributes, the density's influence and significant impact have been highly recorded. Among the environmental degradation and urban sprawl the contagious disease spread leads to new tensions in cities observed from assessment [32]. For population criteria, the weight assigned is performed by potential approaches which describe the COVID-19 spatial distribution and however the temporal variation has not been considered as a drawback. The uniform infection rates have not recorded the COVID-19 transmission

Population-related variables

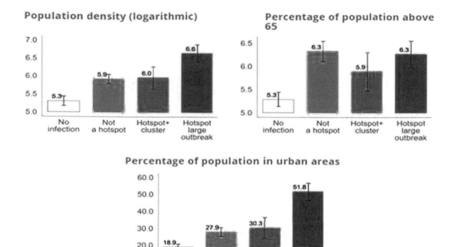

Figure 4.2   Population variables in India.

Source: (Scroll. in/National Family health survey data).

dynamics. The standard model SEIR has been used and does not measure the temporal variation. This study focused on the Brazilian health care system to take an account of the infected patients count. If the control strategies have been affected the infection rate of the long term due to the unclear findings [33]. This study major aim shows the infection or death rates have not been predicted before or disease evolution. At-risk populations have been highly focused and the non-hotspot district characteristics have been analyzed. However, from the below graphical representation, Figure 4.1, shows that the districts with no infections are mostly the rural areas. For denser areas in India, the COVID-19 present burden is higher which the urban areas are usually. For this critical illness, the older people show a larger share of risks [34].

This study developed the contact tracing app in the Netherlands and the dynamics are not considered. The potential uptake alone is predicted from the contact tracing app. For this app promotion, the government and local health authorities put a lot of effort. Personal data sharing has increased due to this app and the respondents may change in the future as the disease risks eased [35]. This study utilized the long term climatic records of population density (PD), air temperature (T), specific humidity (SH), rainfall (R), and wind speed (WS) with topographic altitude (E), actual Evapotranspiration (AET), and solar radiation (SR) at the regional level for the spatial relation association with COVID-19 infection count. With the number of infected cases in India of 36 provinces, the vicariate analysis shows failure in identifying the important relation. The higher importance has been identified by the partial least square technique. After the analysis of various parameters, the present study focused on India shows the COVID-19 infections are more prone to the hot and dry regions with below altitude [36]. The health population is highly infected by the asymptomatic, symptomatic, and pre-symptomatic persons. Another study depicted that the population of asymptomatic patients is higher compared with symptomatic patients. This study has been conducted in India and the improved SIERD model is utilized to predict both kinds of infectious persons. The asymptomatic infected population dynamics were evaluated and this study suggested by making these persons quarantined the number of symptomatic persons also reduced [37].

4.4 Various Deep Learning Model

Promising results obtained from the highly challenging state of art methods related to deep learning. The features interpretation and minimal neural architecture is the challenging one. Various deep learning models like CNN, R-CNN, adversarial models, generative and attention based models have been analyzed in this study. For image segmentation, various analyses and strong research directions have been estimated [38]. For COVID-19 infection prediction, the deep learning models were found to be the most appropriate ones, according to this study. The personal risk scores from lab assessment assigned for the scarce healthcare resources. From this study, the healthcare resource prioritization improved and patient care has been further informed [39]. For predicting the COVID-19 cases of positive this research proposed the deep learning models. State wise comparison has been made based on mild, moderate, and severe in COVID cases. In 32 states, the bi directional LSTM, deep LSTM, and convolutional LSTM have been used for an efficient prediction in which maximum accuracy and absolute error have been chosen. Bidirectional LSTM shows better results. For a short-term prediction, 1 to 3 days BI-LSTM shows better results and it is available publicly. For handling the medical infrastructure these predictions are very helpful for the health authorities. This proposed model can be applied to all nations worldwide [40]. Based on chest X-ray images, three deep CNN approaches have been utilized for COVID-19 detection. With various kernel functions, deep CNN with SVM classifier has been associated. The results depicted asthis study outperformed the local existing approaches. Compared with deep feature extraction, fine-tuning and end to end training needs higher time. The cubic kernel function shows superior performance. Usually, the ResNet-50 model shows better results related to the CNN pre-trained model. Deep CNN performs better for the end to end training process. For the COVID-19 detection, more number of chest X-ray images can be evaluated in future and the various evolution stages can be analyzed to help the radiologists in prediction [41]. This research also utilized the chest radiography images for an efficient COVID-19 prediction by the deep learning approaches. New Coronet model was developed in this study is considered to be low cost and better results obtained. Higher sensitivity and accuracy resulted and thus this model is highly beneficial for the medical practitioners for proper understanding [42].

4.4.1 LSTM Model

In the public health system, strategic planning has been required to avoid deaths from COVID-19. The time series prediction of COVID-19 cases has been performed by LSTM, Bi-LSTM, autoregressive integrated moving average- ARIMA and support vector regression- SVR in 10 major COVID affected countries. This study was estimated by means of the root mean square error, r2-score indices, and absolute error. In this study, BI-LSTM outperforms the other algorithms and it obtains reduced RMSE and MAE values. For better planning and management Bi-LSTM has been considered as a better pandemic prediction algorithm [42]. This work utilized the Canadian health authority and John Hopkins university public datasets for the COVID-19 forecasting model based on deep learning models. For future COVID-19 cases forecasting, this study used the long short term memory- LSTM. The possible ending point of the COVID outbreak was predicted in this study as of June 2020 and compared it with USA, Italy, and Canada transmission rates [43]. Due to the rapid population growth, automatic disease detection is considered a challenging one. However automatic disease detection can support doctors in diagnostics. LSTM is combined with CNN in this study and utilizes the X-ray images to automatically detect COVID-19.

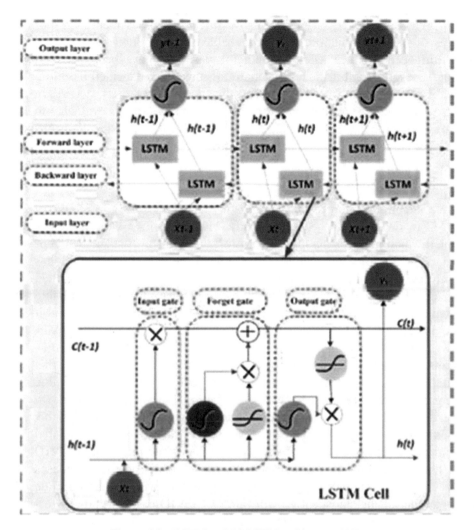

Figure 4.3 LSTM and Bi-LSTM architecture [42].

Better accuracy, sensitivity, specificity have resulted from this proposed system. Rapid diagnosis by doctors has been made from this study [44].

The time series prediction contains the data that is iteratively obtained by the LSTM model. More accurate outputs have been predicted and a number of positive cases have been reported by LSTM. Apart from Google trends data, other data sources can be combined like mass media, screening registers, social media information, environmental and climate factors.

Global prediction is necessary for terms of time series assessment [45]. For a variety of disease prediction, SEIR models have been applied and however, over fitting occurs since a lot of predictor variables have been used. In this study, several combinations of techniques have been executed based on LSTM, XGBoost, and K-means to forecast the short-term COVID-19 cases in the USA. Among the past days and forecasting, the similarity is evaluated in this study using the k means algorithm with the XGBoost technique. K-means with LSTM show larger accuracy as result [46].

4.4.2 Bidirectional LSTM

With the intention of forecasting the cluster data based on the COVID-19 Bi-LSTM model is established in this study. The prediction performance has improved which includes the lockdown information also [29]. The hospitalization estimation for the coming week compared with the present week has been inferred by the four recurrent neural networks. Higher accuracy resulted in predicting the hospitalization in which every patient must receive suitable treatment. The hospitalization requirement has been predicted before and it has the potential to send warning messages to the medical providers [47]. Various tweets have been found worldwide regarding COVID-19 and these kinds of tweets carry valuable information. It is highly challenging to process this information. To analyze the informative tweets, Bi-LSTM and other machine learning approaches are utilized for classification [48]. Various lockdown policies impact with respect to COVID-19 are evaluated and predicted in this study using deep learning techniques. Various scenarios are evaluated related to lockdown policies and their effects are assessed while predicting COVID cases. The lifting of the lockdown especially for schools resulted in increases in infected cases simultaneously [49]. This research provided an appropriate understanding of the statistical growth rate of COVID cases in India. Most affected cases have been predicted using deep learning models [50].

4.5 Conclusion

COVID-19 is the major reason for infecting billions of people and affecting the economy worldwide. This study presented a detailed view of prediction and forecasting the COVID-19 cases worldwide. The forecasting models comprised of various deep learning models such as support vector regression (SVR), long shot term memory (LSTM), bidirectional long short term memory (Bi-LSTM) are assessed for time series prediction of confirmed cases, deaths, and recoveries in ten major countries affected due to COVID-19. The paper also reviewed a deep learning model to forecast the range of increase in COVID-19 infected cases in future days. A comparative study was also performed regarding the discussed deep learning models for COVID-19 prediction. This study provided the guidelines to the various other researchers who focused on the deep learning models in COVID-19 forecasting and prediction.

4.6 Acknowledgement

First and foremost I am extremely grateful to my research mentor, Dr. Y. Vijayalata for her invaluable advice, continuous support, and patience during our book chapter research study. Her immense knowledge and plentiful experience have encouraged me and my teammates in all the time of our academic research and daily lives. We would also like to thank Dr. Siddhartha Ghosh for his technical support and mentorship on our study. It is their kind help and support that have made our book chapter research study worthwhile. Without their tremendous understanding and encouragement in the past few years, it would be impossible for us to complete this study.

4.7 Figures and Tables Caption List

In this book chapter, Section 4.2 describes the deep learning models against covid-19 and their applications, Section 4.3 describes population attributes of COVID-19, followed by section 4.4 describes the various deep learning models and the involved COVID-19 dataset. Finally, the conclusion is presented in Section 4.5.

Figure 4.1 IoT merits towards COVID-19
Figure 4.2 Population variables in India
Figure 4.3 LSTM and Bi-LSTM architecture
Table 4.1: Comparative study of the prevailing literatures

References

[1] B. Xu *et al.*, "Open access epidemiological data from the COVID-19 outbreak," *The Lancet Infectious Diseases,* 2020.
[2] W. Zhang, "Imaging changes of severe COVID-19 pneumonia in advanced stage," *Intensive care medicine,* pp. 1–3, 2020.

[3] S. Arik *et al.*, "Interpretable Sequence Learning for COVID-19 Forecasting," *Advances in Neural Information Processing Systems,* vol. 33, 2020.

[4] P. Nadella, A. Swaminathan, and S. Subramanian, "Forecasting efforts from prior epidemics and COVID-19 predictions," *European journal of epidemiology,* vol. 35, no. 8, pp. 727-729, 2020.

[5] F. Shan *et al.*, "Lung infection quantification of covid-19 in ct images with deep learning," *arXiv preprint arXiv:2003.04655,* 2020.

[6] L. Brunese, F. Mercaldo, A. Reginelli, and A. Santone, "Explainable deep learning for pulmonary disease and corona virus COVID-19 detection from X-rays," *Computer Methods and Programs in Biomedicine,* vol. 196, p. 105608, 2020.

[7] J. Zhang *et al.*, "Viral pneumonia screening on chest X-ray images using confidence-aware anomaly detection," *arXiv preprint arXiv:2003.12338,* 2020.

[8] L. Huang *et al.*, "Serial quantitative chest ct assessment of covid-19: Deep-learning approach," *Radiology: Cardiothoracic Imaging,* vol. 2, no. 2, p. e200075, 2020.

[9] K. Akrawinthawong, K. Majkut, S. Ferreira, and A. Mehdirad, "VOLTAGE-DEPENDENT INAPPROPRIATE RIGHT VENTRICULAR CAPTURE BY RIGHT ATRIAL LEAD PACING AS A CAUSE OF CARDIAC RESYNCHRONIZATION THERAPY NON-RESPONDER," *Journal of the American College of Cardiology,* vol. 69, no. 11S, pp. 2138–2138, 2017.

[10] M. J. Horry *et al.*, "COVID-19 detection through transfer learning using multimodal imaging data," *IEEE Access,* vol. 8, pp. 149808–149824, 2020.

[11] J. Civit-Masot, F. Luna-Perejón, M. Domínguez Morales, and A. Civit, "Deep learning system for COVID-19 diagnosis aid using X-ray pulmonary images," *Applied Sciences,* vol. 10, no. 13, p. 4640, 2020.

[12] M. H. D. M. Ribeiro, R. G. da Silva, V. C. Mariani, and L. dos Santos Coelho, "Short-term forecasting COVID-19 cumulative confirmed cases: Perspectives for Brazil," *Chaos, Solitons & Fractals,* p. 109853, 2020.

[13] T. Chakraborty and I. Ghosh, "Real-time forecasts and risk assessment of novel corona virus (COVID-19) cases: A data-driven analysis," *Chaos, Solitons & Fractals,* p. 109850, 2020.

[14] M. Kamal, A. Aljohani, and E. Alanazi, "IoT meets COVID-19: Status, Challenges, and Opportunities," *arXiv preprint arXiv:2007.12268,* 2020.

[15] A. Rahman, M. S. Hossain, N. A. Alrajeh, and F. Alsolami, "Adversarial examples–security threats to COVID-19 deep learning systems in medical IoT devices," *IEEE Internet of Things Journal,* 2020.

[16] X. Ma *et al.*, "A survey on deep learning empowered IoT applications," *IEEE Access,* vol. 7, pp. 181721–181732, 2019.

[17] H. Jelodar, Y. Wang, R. Orji, and H. Huang, "Deep sentiment classification and topic discovery on novel corona virus or covid-19 online discussions: Nlp using lstm recurrent neural network approach," *arXiv preprint arXiv:2004.11695,* 2020.

[18] A. S. Imran, S. M. Doudpota, Z. Kastrati, and R. Bhatra, "Cross-Cultural Polarity and Emotion Detection Using Sentiment Analysis and Deep Learning–a Case Study on COVID-19," *arXiv preprint arXiv:2008.10031,* 2020.

[19] M. Haider Syed, S. Khan, M. Raza Rabbani, and Y. E. Thalassinos, *"An artificial intelligence and NLP based Islamic FinTech model combining Zakat and Qardh-Al-Hasan for countering the adverse impact of COVID-19 on SMEs and individuals,"* 2020.

[20] Y. Guo *et al.*, "Innovating Computational Biology and Intelligent Medicine: ICIBM 2019 Special Issue," *ed: Multidisciplinary Digital Publishing Institute,* 2020.

[21] E. E.-D. Hemdan, M. A. Shouman, and M. E. Karar, "Covidx-net: A framework of deep learning classifiers to diagnose covid-19 in X-ray images," *arXiv preprint arXiv:2003.11055,* 2020.

[22] Q. Ni *et al.*, "A deep learning approach to characterize 2019 corona virus disease (COVID-19) pneumonia in chest CT images," *European radiology,* vol. 30, no. 12, pp. 6517–6527, 2020.

[23] I. D. Apostolopoulos, S. I. Aznaouridis, and M. A. Tzani, "Extracting possibly representative COVID-19 Biomarkers from X-Ray images with Deep Learning approach and image data related to Pulmonary Diseases," *Journal of Medical and Biological Engineering,* p. 1, 2020.

[24] T. Kufel, "ARIMA-based forecasting of the dynamics of confirmed Covid-19 cases for selected European countries," *Equilibrium. Quarterly Journal of Economics and Economic Policy,* vol. 15, no. 2, pp. 181–204, 2020.

[25] M. Kolhar, F. Al-Turjman, A. Alameen, and M. M. Abualhaj, "A three layered decentralized IoT biometric architecture for city lockdown during COVID-19 outbreak," *IEEE Access,* vol. 8, pp. 163608–163617, 2020.

[26] J. Wang, H. Anh, F. Manion, M. Rouhizadeh, and Y. Zhang, "COVID-19 SignSym–A fast adaptation of general clinical NLP tools to identify and normalize COVID-19 signs and symptoms to OMOP common data model," *ArXiv,* 2020.

[27] B. Tang, Z. Pan, K. Yin, and A. Khateeb, "Recent advances of deep learning in bioinformatics and computational biology," *Frontiers in genetics,* vol. 10, p. 214, 2019.

[28] H. Yi, S. T. Ng, A. Farwin, A. Pei Ting Low, C. M. Chang, and J. Lim, "Health equity considerations in COVID-19: geospatial network analysis of the COVID-19 outbreak in the migrant population in Singapore," *Journal of Travel Medicine,* 2020.

[29] A. B. Said, A. Erradi, H. Aly, and A. Mohamed, "Predicting COVID-19 cases using Bidirectional LSTM on multivariate time series," *arXiv preprint arXiv:2009.12325,* 2020.

[30] Y. Wang, M. Hu, Q. Li, X.-P. Zhang, G. Zhai, and N. Yao, "Abnormal respiratory patterns classifier may contribute to large-scale screening of people infected with COVID-19 in an accurate and unobtrusive manner," *arXiv preprint arXiv:2002.05534,* 2020.

[31] P. C. Silva, P. V. Batista, H. S. Lima, M. A. Alves, F. G. Guimarães, and R. C. Silva, "COVID-ABS: An agent-based model of COVID-19 epidemic to simulate health and economic effects of social distancing interventions," *Chaos, Solitons & Fractals,* vol. 139, p. 110088, 2020.

[32] N. Barak, U. Sommer, and N. Mualam, "Political Environment Aspects of COVID-19: Political Urban Attributes, Density and Compliance," *Density and Compliance (September 07, 2020),* 2020.

[33] W. J. Requia, E. K. Kondo, M. D. Adams, D. R. Gold, and C. J. Struchiner, "Risk of the Brazilian health care system over 5572 municipalities to exceed health care capacity due to the 2019 novel corona virus (COVID-19)," *Science of the Total Environment,* p. 139144, 2020.

[34] A. Clark *et al.*, "Global, regional, and national estimates of the population at increased risk of severe COVID-19 due to underlying health conditions in 2020: a modelling study," *The Lancet Global Health,* vol. 8, no. 8, pp. e1003–e1017, 2020.

[35] M. Jonker, E. de Bekker-Grob, J. Veldwijk, L. Goossens, S. Bour, and M. Rutten-Van Mölken, "COVID-19 Contact Tracing Apps: Predicted Uptake in the Netherlands Based on a Discrete Choice Experiment," *JMIR mHealth and uHealth,* vol. 8, no. 10, p. e20741, 2020.

[36] A. Gupta, S. Banerjee, and S. Das, "Significance of geographical factors to the COVID-19 outbreak in India," *Modeling earth systems and environment,* vol. 6, no. 4, pp. 2645–2653, 2020.

[37] S. Chatterjee, A. Sarkar, M. Karmakar, S. Chatterjee, and R. Paul, "How the asymptomatic population is influencing the COVID-19 outbreak in India?," *arXiv preprint arXiv:2006.03034,* 2020.

[38] S. Minaee, Y. Boykov, F. Porikli, A. Plaza, N. Kehtarnavaz, and D. Ter-zopoulos, "Image segmentation using deep learning: A survey," *arXiv preprint arXiv:2001.05566,* 2020.

[39] T. B. Alakus and I. Turkoglu, "Comparison of deep learning approaches to predict covid-19 infection," *Chaos, Solitons & Fractals,* vol. 140, p. 110120, 2020.

[40] P. Arora, H. Kumar, and B. K. Panigrahi, "Prediction and analysis of COVID-19 positive cases using deep learning models: A descriptive case study of India," *Chaos, Solitons & Fractals,* vol. 139, p. 110017, 2020.

[41] A. M. Ismael and A. Şeng᾿ur, "Deep learning approaches for COVID-19 detection based on chest X-ray images," *Expert Systems with Applications,* vol. 164, p. 114054, 2020.

[42] A. I. Khan, J. L. Shah, and M. M. Bhat, "Coronet: A deep neural network for detection and diagnosis of COVID-19 from chest X-ray images," *Computer Methods and Programs in Biomedicine,* p. 105581, 2020.

[43] V. K. R. Chimmula and L. Zhang, "Time series forecasting of COVID-19 transmission in Canada using LSTM networks," *Chaos, Solitons & Fractals,* p. 109864, 2020.

[44] M. Z. Islam, M. M. Islam, and A. Asraf, "A combined deep CNN-LSTM network for the detection of novel corona virus (COVID-19) using X-ray images," *Informatics in Medicine Unlocked,* vol. 20, p. 100412, 2020.

[45] S. M. Ayyoubzadeh, S. M. Ayyoubzadeh, H. Zahedi, M. Ahmadi, and S. R. N. Kalhori, "Predicting COVID-19 incidence through analysis of google trends data in iran: data mining and deep learning pilot study," *JMIR Public Health and Surveillance,* vol. 6, no. 2, p. e18828, 2020.

[46] S. R. Vadyala, S. N. Betgeri, E. A. Sherer, and A. Amritphale, "Prediction of the number of covid-19 confirmed cases based on k-means-lstm," *arXiv preprint arXiv:2006.14752,* 2020.

[47] Y. Meng, Y. Zhao, and Z. Li, "An early prediction of covid-19 associated hospitalization surge using deep learning approach," *arXiv preprint arXiv:2009.08093,* 2020.

[48] S. Chanda, E. Nandy, and S. Pal, "IRLab@ IITBHU at WNUT-2020 Task 2: Identification of informative COVID-19 English Tweets using BERT," in *Proceedings of the Sixth Workshop on Noisy User-generated Text (W-NUT 2020),* 2020, pp. 399–403.

[49] A. B. Said, A. Erradi, H. Aly, and A. Mohamed, "A deep-learning model for evaluating and predicting the impact of lockdown policies on COVID-19 cases," *arXiv preprint arXiv:2009.05481,* 2020.

[50] A. Dutta, A. Gupta, and F. H. Khan, *"COVID-19: Detailed Analytics & Predictive Modelling using Deep Learning,"* 2020.

5

3D Smartlearning Using Machine Learning Technique

M. Srilatha[1], D. Nagajyothi[2], S. Harini[3], and T. Sushanth[4]

[1]Assistant Professor, Department of ECE, Vardhaman College
of Engineering, India
[2]Associate Professor, Department of ECE, Vardhaman College
of Engineering, India
[3,4]Student, Department of ECE, Vardhaman College of Engineering, India
E-mail: m.srilatha@vardhaman.org; d.nagajyothi@vardhaman.org;
sayiniharini2k@gmail.com; thadishetti.sushanth@gmail.com

Abstract

One of the most significant areas of image processing and computer vision is object detection. In modern days, Humans have a tendency to grasp things fast when we teach them in a practical way. The proposed system is designed to learn the correspondence between preached words and conceptual visual attributes from a spoken image description dataset. For this purpose, vision technology and neural network algorithm are used to do image enhancement and manipulation techniques using LABVIEW platform. First, train the PC with OCR (Optical Character Recognition) technique and continue this process for 2-3 times so that it becomes accurate. Then with the help of the KNN algorithm, an input image is compared with the predefined dataset. The acquisition and processing of images are done in the graphical programming environment of LabVIEW. This gives all the benefits of this software to the application: modularity, effortless realization, desirable user interface, springiness, and the ability to develop very simple new features. The learner can use the proposed scheme to learn things in a smarter way.

Keywords: LabVIEW, myDAQ, Vision and Motion, Vision acquisition, Optical Character Recognition, Machine Learning, KNN algorithm, Smart Learning, 3D Technology

5.1 Introduction

The way quality control processes are conducted has been transformed by 3D imaging, which provides all dimensional data to users. In general, the physical environment is three-dimensional in nature which has width, height, and depth, and we passage around every day in 3D, while a traditional 2D image has width and height, but it does not technically have depth. 3D imaging [1] is a technique in which an image develops or creates the illusion of depth. 3D imaging has become a very useful factor to assist in quality control processes for industrial applications.

In order to create a 3D interpretation for inspection and testing purposes, many different technologies can help with this process. For quality control purposes, a 3D image will offer users a realistic replica of the object. Some of the most common 3D imaging applications are in the aerospace, automotive, and medical device industries. 3D imaging allows users to replicate and analyze, in full 3D form, parts, and objects. Using the rules of perspective, the computer maps virtual 3D objects into 2D screen space. It is desirable to simulate as many of the perception tools as possible in order to represent the 3D world on a flat surface such as a display screen. When interactive 3D images are made, the user can feel and engaged with the scene, this experience is called virtual reality.

In today's digital age, Smart Learning [2] is a comprehensive term for education. It reflects how advanced technologies allow learners to more realistically, competently, and conveniently grasp knowledge and skills. Rather than a static follower of the educational process, the learner becomes a proactive leader. Smart learning aims to provide students with holistic learning using modern technology to prepare them fully for a fast-changing world in which adaptability is essential.

A paradigm shift in the way students access education is offered by smart education. Smart learning introduces students to a progressive and natural methodology that develops the subject from 0 percent to 100 percent with full phonological immersion.

Technology has advanced to an amazing speed, which also raises the need for it as well. The increased demand for technology can be seen in different forms that can do many things, particularly helping to needs for humans. An example of technology's growth is machine learning.

Machine learning [3, 4] is a type of artificial intelligence that enables accurately the software applications in forecasting outcomes without being specifically programmed to do so. Machine learning algorithms use chronological data as input in order to forecast new output values. There are many enhancements in the field of machine learning and identifying the character is one of the budding active feilds in which machine learning can be used extensively.

The main goal is to make use of machine learning algorithms for children to learn things in a smarter way and in 3D mode. Children grasp things fast when the teaching has happened in a practical way. The proposed idea will help children to learn colors and shapes in a smatter way without any stress.

The proposed method is implemented using LabVIEW [5] software which is a graphical programming language and is easy to implement and understandable. It is a user-friendly language. In this proposed design video will be acquired through webcam and crop the required portion which becomes an input to the sub functions to find the color and shape of the cropped image. After certain training, compared with the template images, finally the output is observed in the form of voice.

Using this design, learners will not feel a burden to study, they keep more interest in learning as it is like playing the game. This design has low complexity, easily understand by new learners, and implemented at a low cost.

A remainder of the chapter is as follows. Chapter 2 provides insight into existing methods from a brief survey. The methodology used in the proposed system is explained in Chapter 3. Chapter 4 demonstrates the proposed method results and discussion followed by a conclusion in Chapter 5.

5.1.1 Literature Survey

5.1.1.1 Machine learning basics

Machine learning [3, 35] is a type of artificial intelligence that enables accurately the software applications in forecasting outcomes without being specifically programmed to do so. Machine learning algorithms use chronological data as input in order to forecast new output values. There are many enhancements in the field of machine learning and identifying the character is one of the budding active fields in which machine learning can be used extensively.

Machine Learning Algorithms are basically divided into four types: Supervised, Unsupervised, Semi supervised, and Reinforcement.

5.1.1.1.1 Supervised learning

In this [37, 38], classification is done manually which becomes easy for computers. It is like giving a standard answer to a computer and based on this the computer will reply. The reliability of this method is more since the training data will contain inputs with a standard answers. In general, the supervised algorithm is written as

$$Z = f(y) \tag{5.1}$$

Where Z is the predicted output and y is the input and the function is used to map input features with standard output.

5.1.1.1.2 Unsupervised Learning

Unsupervised learning [39] is applied to input data that will not come under either classification or labeled. Since manual classification is not used so it is easy for humans. To the computer, standard answers will not be given. It gives the output depends on examined information. This algorithm is mostly used in data mining.

5.1.1.1.3 Semi supervised learning

It is a combination of supervised and unsupervised learning [40, 41] where it can be applied to data that is partially labeled. An example of this learning is a speech analysis.

5.1.1.1.4 Reinforcement learning

This algorithm is applied to the system which can interact with its surroundings and depends on this errors and rewards will be calculated.

Author [10, 11] proposed a new method to learn the words by drawing images. They developed a kit in which the relation between words and audio is predefined. At first, training is done by using GAN and CLEVERGAN (a new dataset of audio descriptions of GAN-generated CLEVR images) and then training is done in some specific areas which differ in an informative way.

Author [12, 13] has given a study on various teaching pattern strategies using specific mathematical analysis methods. Using this studyimproves smart learning and enhances ETIAS media coding. He also suggested that even more improvement can be done if mixed methods are used.

The learning [14] can also be done by identifying shape and appearance using two non-neighboring features which are side-inner difference features (SIDF), that can differentiate between the background and pedestrian and the difference between the pedestrian contour and its inner part can be found by symmetrical similarity features (SSF).

Author [15] has shown that pedestrians can also be identified using 3D points and color images. For this purpose, a camera and a 3D sensor is used to increase the efficiency. This proposed design adopts DBSCAN to cluster 3D points and to project these clusters onto their selected region of interest. Due to this, its performance is increased because it focuses on ROI instead of the whole image. As the small objects have less resolution and shape is not clear, so identification of these objects has been ignored.

So, author [16] proposed a new method to overcome this problem. He suggested an up-sampling method that gives better performance than the current state-of-the-art for end-to-end small object detection. Generally, Object detection of moving objects is complex when compared to stationary objects.

To identify the objects in a video, the author [17] captured frames and pictures is given to the algorithm. There may be a case in which 2 or more frames to be captured for the identification of one object. The main goal is to make minimum delay between the capturing of two frames.

Author [18, 19] tries to explain the differences between object tracking and object detection and concludes by saying that object tracking is more accurate. They try to combine both and make detection more efficient in fast videos. A new technique named Computational expensive detection network is applied in this process. The author says that non-key frames are applied when the previous key fails for detection.

Author [20] designed a system that will acquire the image and display the output in real time. For this purpose, the author used a CMOS sensor to acquire the image and an HDMI device to display the output and to control input and output devices, an FPGA is used.

Author [23] tries to identify the color and the properties of soil using digital image processing techniques. The more fertile the soil the more it is efficient. Generally, a chart named Munsell soil color chart is used to assess the efficiency of soil.

A key topic in Intelligent Transportation is to determine the vehicle density happening on the road. Reliable and robust video sequence vehicle detection is an important issue in traffic control applications. In real time, the vehicle detection system's processing time should be noted. If the traffic is high, widening of the roads should be done or the diversion to other routes should have happened so that the next time traffic jam will not occur.

The system's proposed algorithm [24] detects and tracks vehicles in real time. Vehicle detection is only performed on the vehicles' color characteristics. Using an enhanced Kalman filter method, they are tracked after the vehicles are detected. The data association cost matrix plays a significant role in assigning a centroid to each enhanced Kalman filter. So, that filter from Kalman is tracking the same centroid.

Author [25] uses an enhanced image color cast detection algorithm that incorporates the parameter of the color cast function into the method of the 2D Lab histogram. It also makes the image classification of the color deviation more perfect and optimizes algorithm flow.

This enhanced algorithm can achieve greater detection accuracy and efficiency compared with the 2D Lab histogram method. Face detection is one the most developing technique from the past few years. As technology is increasing, human skin detection is playing a vital role. There may be some cases in which skin is not detected due to low-light conditions, legacy grayscale images and videos, and near-infrared images. So, author [26] focused on gray scale images and used their texture and pixel value for detection of the skin tone irrespective of color.

Author [27] has proved that robot cars can move in different directions and navigate based on color detection. Robots can also be controlled using voice. The technology for voice recognition allows the system to understand and comply with the voice command and to analyze man-made languages. While there are several touch-based robots and other control-aiding devices, control over voice with simple operation is left behind.

The primary objective [30] is to create a gateway for voice-controlled robot automation for easy operation. As a result, a voice-controlled robot is designed by using speech recognition technique and LabVIEW. Compared to existing systems, the proposed system with LABVIEW makes research more comfortable and simpler. Voice conversion (VC) can be used to modify speech characteristics, such as the identities of gender and speakers.

The VC can be applied to different tasks, such as speaking assistance and anonymization of speakers. These VC methods generally require parallel speech data for training, which is very costly. A new technique called Cycle-GAN [31] has recently been implemented for voice conversion, which does not require parallel speech data. In this paper, the author further expands the concept of using CycleGAN to convert the voices of multiple speakers by conditioning the CycleGAN using information about speaker identity.

Author [32] proposed a framework that extracts a set of interpretations that are equivalent to the seized object shape and structure. To achieve this, the author used a technique called shape- collapsing method which is used through a supervised minimization of energy. This framework is adaptable to find out a way to detect existing CityGML levels and levels given by experts. Extensive outdoor scenes can be converted into digital using the LiDAR scanning technique. But the difficulties faced by the LiDAR technique are, some regions may miss or noise may get added with the data.

To overcome this problem, author [33] proposes a system where the scanning will be happened locally by using adaptive kernel-scale scoring. A significant explanation for deaths among older people correlates to accidental falls. These falls occur due to lack of medicalsince they live alone and unable to get help during accidents. To solve this problem, author [34] proposed a system to identify fall detection.

Finally, to more accurately determine whether a fall event has occurred or not, two additional features, acceleration, and angular acceleration, are proposed. Research work of various authors is listed in the Table 5.1.

Table 5.1   Research work of various authors

Ref. No.	Proposed Method	Features	Outcome
[10]	CLEVRGAN	Color and Size	• Supervision other than spoken captions and GAN is required • Improvises the learning ability • With the proposed framework, words can be visualized in terms of color and size
[12]	Lag Sequential Analysis tool in particular General Sequential Querier is used	–	• In a smart Learning Environment, various teacher behavioral patterns can be recognized • Various promotion approaches can be known for teaching actions

Continued

Table 5.1   (*Continued*)

Ref. No.	Proposed Method	Features	Outcome
[14]	Non-neighboring features: Side-inner difference features and Symmetrical-similarity features	Appearance and Shape	Using Non-neighboring features for detection, the average miss rate is decreased by 4.44%
[15]	DBSCAN and SVM classifier	3D points and color	The proposed algorithm will detect only Region Of Interest, so the technique is more efficient
[16]	Faster R-CNN	FlickrLogos dataset	Small objects can also be detected accurately
[23]	K-Nearest Neighbor	Color	Soil color can be detected efficiently and also can easily seperate the soil part from the background
[24]	Enhanced Kalman Filter	Color	All the vehicles in the video will be calculated and based on the number of vehicles on the road, the algorithm can decide whether the road is crowded or not
[26]	Region growing algorithm	Gray scale image	Without the use of color, skin detection has happened automatically
[30]	Speech Recognition Technique	Voice	Based on the human voice, the robot will respond automatically
[31]	CycleGAN	Voice	The proposed method will transform more than one speaker's voice into another speaker's voice
[32]	Parameter-free algorithm	Shape	The structural scales can be reliably detected, and their corresponding characteristics can be learned by using training sets of different kinds of man-made objects

5.2 Methodology

This section emphasis on the proposed method and also discuss the proposed approach for two different cases. First, the proposed system block diagram is elaborated and then the proposed approach along with the flow diagram is explained for stationary objects and also moving objects.

5.2.1 Problem Definition

In our day-to-day life, we see many children committing suicide due to a lack of scores in their academics. This is because they are stressed and are fed up with their curriculum. To overcome this problem and to relief their stress to some extent here we are proposing a new technique of learning called smart learning through LabVIEW [28, 29]. This design helps children to identify the colors and shapes of the objects and they enjoy learning.

5.2.2 Block Diagram of Proposed System

Figure 5.1 illustrates the proposed system block diagram.

5.2.2.1 myDAQ

In the myDAQ student data acquisition device, as shown in Figure 5.2, has 8 universally plug-and-play virtual instruments built on LabVIEW like function generator, digital multimeter, and oscilloscope. In combination with LabVIEW and NI Multisim software, myDAQ allows real engineering and gives students the power to prototype systems and analyze circuits beyond conventional lectures and labs. Pin description of myDAQ is shown in Table 5.2.

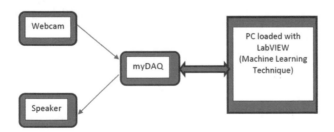

Figure 5.1   Proposed system block diagram.

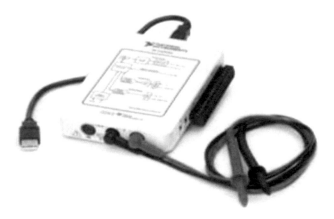

Figure 5.2   NI myDAQ [6].

Table 5.2   Pin description of NI myDAQ [6]

Signal Name	Reference	Direction	Description
AUDIO IN	—	Input	Audio Input—Left and right audio inputs on a stereo connector
AUDIO OUT	—	Output	Audio Output—Left and right audio outputs on a stereo connector
+15V/–15V	AGND	Output	+ 15 V/–15 V power supplies
AGND	—	—	Analog Ground—Reference terminal for AI, AO, +15 V, and –15 V
AO 0/AO 1	AGND	Output	Analog Output Channels O and 1
AI 0+/AI 0–; AI 1+/AI 1	AGND	Input	Analog Input Channels 0 and 1
DIO	DGND	Input or Output	Digital I/O Signals—General-purpose digital lines or counter signals

Table 5.2 (*Continued*)

Signal Name	Reference	Direction	Description
DGND	—	—	Digital Ground— Reference for the DIO lines and the +5 V supply
PFI O/CTR 0 SOURGE	—	—	Digital I/O, line 0; PFI 0, Default function: Counter 0 Source
PFI 1/ CTR 0 GATE	—	—	Digital I/O, line 1; PFI 1, Default function: Counter 0 Gate
PFI 2/CTR O AUX	—	—	Digital I/O, line 2; PFI 2, Default function: Counter 0 Aux
PFI 3/CTR O OUT	—	—	Digital I/O, line 3; PFI 3, Default function: Counter 0 Out

5.2.2.2 Speaker

Computer speakers require a power source externally since it has an internal amplifier. Computer speakers widely vary with respect to price and quality. Computer speakers are small, plastic, and have mediocre sound quality, sometimes packaged with computer systems. There are equalization features such as bass and treble controls in some computer speakers. An Aux jack and a compatible adapter can be used to connect Bluetooth speakers to a computer.

5.2.2.3 Camera

A webcam, as shown in Figure 5.3, is a video camera that feeds or streams a picture or video to a computer network, such as the internet, in real time, or through a computer. Typically, webcams are small cameras that reside on a desk, connect to the monitor of a user, or are built into the hardware. During a video chat session involving two or more people, with conversations including live audio and video, webcams can be used. Webcams are known for their low production costs and high versatility, making them the video telephony type that is most cost-effective. The resolution steadily increased

Figure 5.3   Webcam.

from 320 to 240 to 640 to 480 as webcams developed concurrently with display technologies, USB interface speeds, and internet broadband speeds, and some new versions also offer 1280 to 720 or 1920 to 1080 resolution.

5.2.3 Optical Character Recognition

OCR is an electronic or mechanical translator which translates images of any form like a scanned document, a photograph of a document, a photo of a scene or a subtitle of text to machine-coded text. OCR [8, 9] is a generalized method that converts text or images into printed texts digitallyso that corresponding text can be searched, edited, or compactly processed, viewed online, and used in machine processing electronically. This method is the area of science in computer vision, pattern recognition, and AI. The early versions need to train each character with visuals and applicable to one font at a time only, but the advanced versions are very popularsince they can produce a high degree of recognition for most fonts and support a number of inputs in digital format. OCR system consists of six stages, they are acquisition, segmentation, pre-processing, feature extraction, recognition, and post processing as shown in Figure 5.4.

5.2.3.1 Acquisition

In this stage, the image is acquired from an external device like a camera or tablet or scanner. The acquired image will be converted into machine readable form so that it can be processed further.

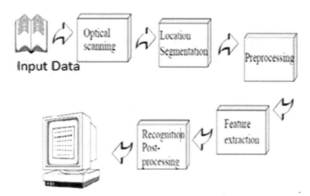

Figure 5.4   Stages of OCR [7].

5.2.3.2 Segmentation
In this stage, the input image is subdivided into segments, so that we know what exactly the input image has, and also processing can be done easily with greater accuracy.

5.2.3.3 Pre-Processing
In this stage, the segmented image will be enhanced to improve the quality. Different types of pre-processing techniques like thresholding, noise removal are used depending on the application.

5.2.3.4 Feature Extraction
In this stage, different features will be extracted to identify the uniqueness of the chracters. Depending on the feature extraction only the accuracy of the classification technique depends. The input image features extracted will fall into three categories: Global Transformation, Structural features, and Statistical features.

5.2.3.5 Recognition
It is also called classification. The efficiency of this stage depends on what parameters had been extracted. In this stage, the parameter or character will be recognized and classified into its respective group.

5.2.3.6 Post-Processing
Various approaches like contextual analysis, geometrical and document context, lexical processing are used to improve the efficiency of the results, once the parameter or character is classified into its respective group.

5.2.4 K-Nearest Neighbors Algorithm

KNN algorithm will be used for two purposes: Classification and Regression. Input will be the same for both, but the output will be decided based on whether we are using KNN [21, 22] for classification or regression. KNN is very simple and can be easily implemented, but the difficulty with this algorithm is if the size of the training set increases then computations become expensive, and also accuracy will decrease if the input features are noise components or irrelevant components. KNN algorithm [36] can be used in various fields like a diagnosis of more than one disease which shows same symptoms, analyzing financial matters before sanctioning loan, recognizing videos and images, casting votes for various parties.

For continuous variables, the distance metric used is Euclidean and for discrete variables, the distance metric is Overlap. The implementation of KNN is done by first transforming the input data set into a feature vector, then it calculates the distance between the points using Euclidean distance which is shown below.

$$d\,(a,b) = \sqrt{(b_1 - a_1)^2 + (b_2 - a_2)^2 + \cdots + (b_k - a_k)^2}$$

$$= \sqrt{\sum_{n=1}^{k} (b_n - a_n)^2} \tag{5.2}$$

Where a is the training data, b is the test data point, d(a,b) is the Euclidean distance between a and b and k is the number of pints in the data. When this formula runs, it will calculate the distance between input data set with test data and discovers the probability of similarity with test data. Now depends on the highest probability, the classification of input data sets will be done. The working of the KNN algorithm is explained with the following steps:

Step 1: Load input data and test data.

Step 2: Choose the value of K.

Step 3: Using the Euclidean metric or any other metric, calculate the distance between each row of input data with each point in test data.

Step 4: Arrange them in ascending order.

Step 5: Select the topmost K row.

Step 6: Based on the probability of similarity, the classification of data sets will happen.

5.2.5 Proposed Approach

Figure 5.5 shows the flow diagram for the proposed system. First the image is acquired from the webcam and is stored in a template. The design is split into two halves. The first half detects the color while the second half detects the shape. In the first half, the image is given to the 'Color learn function' which segregates the saturation values of the image and this output is a range and coerce function. If the values are in range, the output is given to a case structure and in case of structure with the help of invoke node and the property node the text is converted into speech. Similarly, in the second half also same procedure takes place but here the image is given to 'Detect shapes function'.

CASE 1: Stationary Objects

- At first, image is acquired with the help of IMAQ create and IMAQ reading vision and mission software.

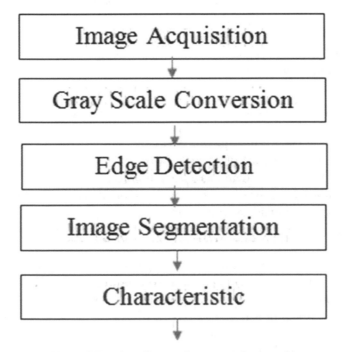

Figure 5.5 Flow diagram for proposed approach.

- The whole program is divided into two parts. 1st part is used to detect the color while the next part is used to detect the shape.
- Here one process should occur after completion of the other process. So, a sequential structure is used in this VI.
- Region of interest is selected with the help of the property node.
- This ROI is given to IMAQ color learn, the color learn function segregates the colors with respect to their saturation values.
- Then this is passed to a 'range or coerce' function which gives output as '1' if they are in range else gives '0'
- All these are combined as an array with the help of 'Built array'.
- With the help of 'search index array' the color is identified and the index is given to a case structure.
- Inside the case structure, a popup message is stored and also with the help of invoke node and property node text is converted into voice
- Similarly, to identify the Shape we extract a single color from the RGB plain.
- Then ROI is selected and is given to 'IMAQ Find pattern'
- If the Shape of the cropped image matches with the template, then it produces an output as '1' else '0'.
- All these are combined using a built array function and are given to a case structure as in the case of identification of color.

CASE 2: Moving Objects

- This design works for moving objects also.
- Video means a sequence of images. So here we try to acquire images continuously and store them in sequential order.
- To continuously acquire images a while loop is used. So, the whole program is placed inside a while loop.
- To do this, IMAQ open and IMAQ grab functions are used to acquire the image.
- The whole program is divided into two parts. 1st part is used to detect the color while the next part is used to detect the shape.
- Here one process should occur after completion of the other process. So, a sequential structure is used in this VI.
- Region of interest is selected with the help of the property node.
- This ROI is given to IMAQcolor learn, the color learn function segregates the colors with respect to their saturation values.

- Then this is passed to an in 'range or coerce' function which gives output as '1' if they are in range else gives '0'
- All these are combined as an array with the help of 'Built array'.
- With the help of 'search index array' the color is identified and the index is given to a case structure.
- Similarly, to identify the Shape we extract a single color from the RGB plain.
- Then ROI is selected and is given to 'IMAQ detect shapes'
- If the Shape of the cropped image matches with the template, then it produces an output as '1' else '0'.
- All these are combined using a built array function and are given to a case structure.
- At the outside of the while loop also, we use a sequential structure, so that color and shape outputs are popped one after the other.
- In the case, structure Invoke node and property node are used to convert the text into voice.

5.2.6 Discussion of Proposed System

5.2.6.1 Flow Chart

Case 1: Stationary Objects

Figure 5.6   Flow of Execution for Stationary Objects.

Case 2: Moving Objects

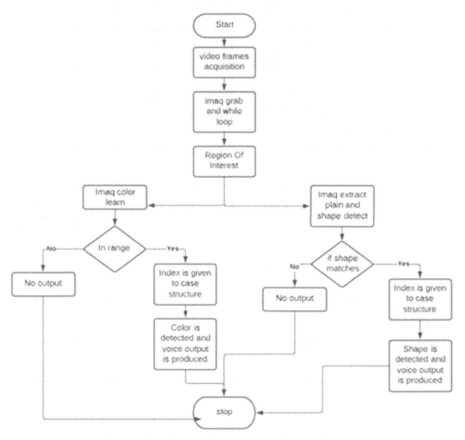

Figure 5.7   Flow of Execution for Moving Objects.

5.2.6.2 Algorithm

Step 1: Select ROI from the image using the property node.

Step 2: Using color learn function segregates all the colors using intensity and saturation values.

Step 3: Use In range and coerce function to find whether it exists within the range or not.

Step 4: Similarly, to detect the shape, First convert RGB to Gray Scale.

Step 5: To convert use the function Extract SInleplane, Now select the ROI from this image and give to detect shape function,

Step 6: You Can also train the figures and compare them with the template with the help of the IMAQ find a pattern.

Step 7: For Training NI Vision Assistant need to be used.

5.3 Results and Discussion

The PC loaded with LabVIEW tool kit along with vision and mission and Image acquisition Software is used to implement the proposed system. The hardware components used are NI myDAQ to connect Audio IN and Audio Out, speaker, and camera. Table 5.3 shows the specifications of the toolkit used an icon description is shown in the Table 5.4.

CASE 1: Stationary Objects

In this case, the input objects are stationary and are applied as input. Figures 5.8 and 5.9 shows how the machine learning algorithm extracts features like identification of color and shape of an object, alphabets, and corresponding block diagram is shown in Figures 5.10 and 5.11 and the

Table 5.3 Specifications of the toolkit

Type of Software	Version
NI Labview Software	2018
Vision and motion	2018
Vision acquisition	2018
Machine learning tool kit	2018

Table 5.4 Information on icons

Function	Description
Imaq Read	To take the image
Imaq create	To store the image
Imaq color learn	To identify the color
In range or coerce	To identify whether they are in given range or not
Build array	To form an array
Search 1D array	To search an element in an array
Case structure	To display all the possible cases
Imaq extract	To convert RGB to black and white
Imaq find pattern	To detect the shape
Imaq open	To acquire video
Imaq grab	To store video in the form of frames
Imaq detect objects	To detect the shape of object
Stop	To end the acquisition

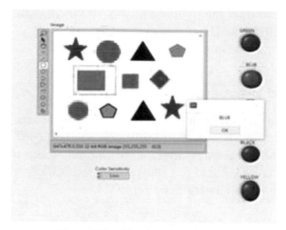

Figure 5.8   Identification of color.

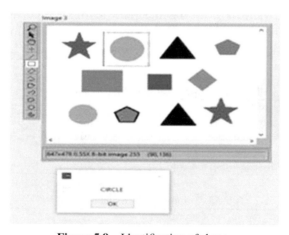

Figure 5.9   Identification of shape.

corresponding things will be displayed on the screen and at the same time audio will have listened from a speaker.

when the image is acquired using the camera, it is stored in image create which is an RGB input. To identify the color, the image is given to image color learn which uses the 'KNN' technique and segregates the intensity values and these values are extracted with the help of index array function. The output of the index array is given to 'In range and Coerce' function which checks whether the values are in range and produces output as '1' if it is true

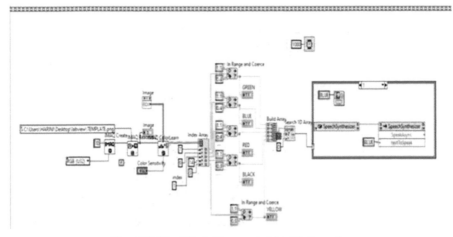

Figure 5.10   Block diagram to identify the color.

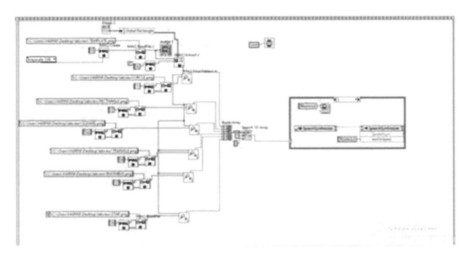

Figure 5.11   Block diagram to identify the shape.

or else '0'. These outputs are combined as an array with the help of the built array function and a 'search 1D array' is used to extract the index of the color which is identified and is given to the case structure. Case structure has a pop-up message inside it and also converts text to voice with the help of invoking node and the property. The process is shown in Figure 5.10. Similarly, to identify the shape we extract a single plane from the image and given to image find a pattern, which compares the input image with the already predefined trained template. Training is done with the help of the OCR technique which is called unsupervised learning (i.e, the more no of times you train the more efficient will be the output) again the output is given to a building array passed to the search 1D array and case structure and the output is obtained as shown in Figure 5.11.

CASE 2: Moving Objects

In this case, the input objects are not stationary and are applied as input. Figure 5.12 and 5.13 shows how machine learning algorithm extracts features like identification of color and shape of a moving object, alphabets, and respective block diagram is shown in Figures 5.14, 5.15 and 5.16 and the corresponding moving thing is captured in real time and will be displayed on the screen and at the same time audio will have listened from the speaker.

Figure 5.12   Identification of color.

Figure 5.13   Identification of shape.

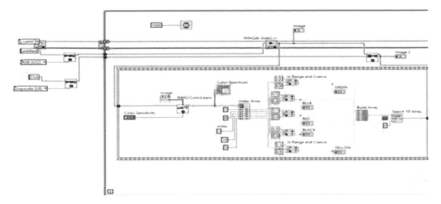

Figure 5.14 Block diagram: Part 1.

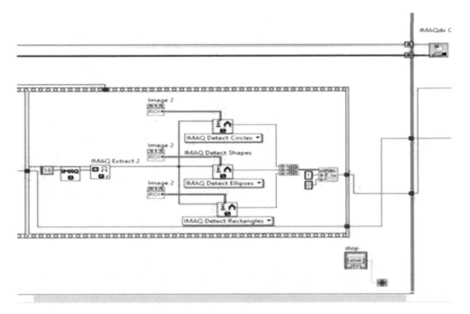

Figure 5.15 Block diagram: Part 2.

This case elaborates on the functioning of moving objects. Here we use a technology called vision acquisition to acquire moving objects or videos. We know that video means a collection of frames. So here a while loop is used to continuously acquire the images and take the single image and follow the same procedure.

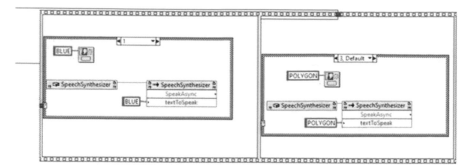

Figure 5.16   Block diagram: Part 3.

5.4 Conclusion and Future Scope

As new technologies are developed and advancements came into the picture, learners can learn more effectively, competently, flexibly, and contentedly. Increased attention has been given to smart education, a concept that defines learning in the digital age. With user-driven and motivational learning solutions, the proposed system aims to promote 21st century learning. The technology advancements and the impact on student learning have been seen when we move from blackboard teaching to smart classes. Similarly, the proposed system also helps children in many ways like self-learning, easily understanding, stress relief, and interest. As the 21st century is declared as an 'Era of Artificial Intelligent' where human power is replaced with robotic technology. In the same way, the proposed system also aims at teaching learners without a mentor effectively and efficiently. The design and implementation of 3D smart learning using LabVIEW are presented in this chapter. Required image is cropped using the ROI and some required characteristics are extracted and are compared with the existing template. When the two images match, the output is displayed and voice output is obtained. The future education system can be performed using the proposed system which becomes an interesting and effective tool to teach learners through online also.

References

[1] https://computer.howstuffworks.com/3dgraphics1.htm
[2] Zhu. ZT, Yu. MH, Peter Riezebos, "A Research Framework of Smart Education", Smart Learning Environment. 3, 4, 2016.

[3] https://searchenterpriseai.techtarget.com/definition/machine-learning
-ML.

[4] Ibtehal Talal Nafea, "Machine Learning in Educational Technology",
Intechopen, Advanced Techniques and Emerging Applications, September 2018, DOI"10.5772/intechopen.72906.

[5] National Instruments (Access at http://www.ni.com).

[6] Data Acquisition hardware-myDAQ (Access at http://ni.cpm/pdf/man
uals/373060g.pdf).

[7] Sarika Pansare, Dhanshree Joshi, "A Survey on Optical Character Recognition Techniques", International Journal of Science and Research, Vol. 3, Issue. 12, pp. 1247–1249, December 2014.

[8] S.K. Singla, R.K. Yadav, "Optical Character Recognition based speech synthesis system using LabVIEW", Journal of Applied Research and Technology, Vol. 12, Issue 5, pp. 919–926, October 2014.

[9] Tapan Kumar Hazra, Dhirendra Pratap Singh, Nikunj Daga, "Optical Character Recognition using KNN on custom image dataset", 8th Annual Industrial Automation and Electromechanical Engineering Conference, August 2017.

[10] DidacSuris, Adria Recasens, David Bau, David Harwath, James Glass, Antonio Torralba, "learning words by drawing images", 2019 IEEE/CVF Conference on Computer Vision and Pattern Recognition, pp. 2774–2784, 2020.

[11] T. Afouras, J. S. Chung, A. Zisserman, "The Conversion: Deep audio-visual speech enhancement". In INTERSPEECH, 2018.

[12] Yu, Luyao, Yulu Cui, and Hai Zhang. "Teacher behavior sequence under smart learning environment." 2018 International Joint Conference on Information, Media and Engineering (ICIME). IEEE, pp. 158–161, 2018.

[13] Garciafarina. A, Jimenezjimenez. F, Anguera, M. T, "Observation of Communication by Physical education teachers: Detecting patterns in verbal behavior", Frontiers in Psychology, 9, 334, 2018.

[14] Jiale Cao, Yanwei Pang, Xuelong Li, "Pedestrian Detection Inspired by Appearance Constancy and Shape Symmetry", 2016 IEEE Conference on Computer Vision and Pattern Recognition, pp. 1316–1324, 2016.

[15] Lin, B. Z., & Lin, C. C, "Pedestrian detection by fusing 3D points and color images", 2016 IEEE/ACIS 15th International Conference on Computer and Information Science (ICIS) (pp. 1–5). IEEE, June 2016.

[16] Eggert Christian, Dan Zecha, Stephan Brehm, Rainer Lienhart, "Improving small object proposals for company logo detection." Proceedings of the 2017 ACM on International Conference on Multimedia Retrieval, pp. 167–174, 2017.

[17] Lao, Dong, and Ganesh Sundaramoorthi. "Minimum delay moving object detection." Proceedings of the IEEE Conference on Computer Vision and Pattern Recognition, pp. 4809–4818, 2017.

[18] Yang Wenfei, BinLiu, Weihai Li, Nenghai Yu, "Tracking assisted faster video object detection." 2019 IEEE International Conference on Multimedia and Expo (ICME). IEEE, pp. 1750–1755, 2019.

[19] Xizhou Zhu, Jifeng Dai, Lu Yuan, Yichen Wei, "Towards high performance video object detection", CVPR, 2018.

[20] Shi, Haitao, and Shiru Zhang. "Dual-channel image acquisition system based on FPGA." 2019 International Conference on Intelligent Transportation, Big Data & Smart City (ICITBS). IEEE, pp. 421–424, 2019.

[21] Xiaoyan Zhu, Chenzhen Ying, Jiayin Wang, Jiaxuan Li, Xin Lai, Guangtao Wang, "Ensemble of ML-KNN for classification algorithm recommendation", Knowledge-Based Systems, Vol. 221, June 2021.

[22] K. Maheswari, A. Priya, A. Balamurgan, S. Ramkumar, "Analyzing student performance factors using KNN Algorithm", Materials Today: Proceedings, February 2021.

[23] ManiyathShima Ramesh, Ramachandra Hebbar, Akshatha K N, Architha L S, S. Rama Subramoniam, "Soil Color Detection Using Knn Classifier." 2018 International Conference on Design Innovations for 3Cs Compute Communicate Control (ICDI3C). IEEE, pp. 52–55, 2018.

[24] Anandhalli, Mallikarjun, and Vishwanath P. Baligar. "Vehicle Detection and Tracking Based on Color Feature." 2017 International Conference on Recent Advances in Electronics and Communication Technology (ICRAECT). IEEE, pp. 240–247, 2017.

[25] Bai, Lefu, et al. "An Improved Image Color Cast Detection Algorithm." 2020 International Conference on Culture-oriented Science & Technology (ICCST). IEEE, 2020.

[26] Sarkar, A., Abbott, A. L., & Doerzaph, Z, "Universal skin detection without color information", In 2017 IEEE Winter Conference on Applications of Computer Vision (WACV) (pp. 20–28). IEEE, March 2017.

[27] Duan, Li, and Zichun Yu. "Robot Cars Navigation by Color Detection." 2018 International Conference on Smart Grid and Electrical Automation (ICSGEA). IEEE, 2018.

[28] Alan S. Morris, Reza Langari, "Chapter 9-Use of LabVIEW in data acquisition and postprocessing of signals", Measurement and Instrumentation (Third Edition)-Theory and Applications, pp. 243–274, 2021.

[29] Ashok Kumar L, Indragandhi V, Uma Maheswari Y, "Chapter 7-Graphical Programming Using LabVIEW for Beginners", Software Tools for the Simulation of Electrical Systems-Theory and Practice, pp. 239–286, 2020.

[30] Patil, Shwetha, and Abhigna Abhigna. "Voice controlled robot using labview." 2018 International Conference on Design Innovations for 3Cs Compute Communicate Control (ICDI3C). IEEE, pp. 80–83, 2018.

[31] Yook, Dongsuk, In-ChulYoo, and SeunghoYoo. "Voice conversion using conditional CycleGAN." 2018 International Conference on Computational Science and Computational Intelligence (CSCI). IEEE, pp. 1460–1461, 2018.

[32] Fang, Hao, Florent Lafarge, and Mathieu Desbrun. "Planar shape detection at structural scales." Proceedings of the IEEE Conference on Computer Vision and Pattern Recognition, pp. 2965–2973, 2018.

[33] Wang, Jun, and Kai Xu. "Shape detection from raw lidar data with subspace modeling." IEEE transactions on visualization and computer graphics, Vol. 23, Issue. 9, pp. 2137–2150, 2016.

[34] Lin, Chih-Yang, Shang-ming wang, Jia-Wei Hong, Li-Wei Kang, Chung-Lin Huang, "Vision-based fall detection through shape features." 2016 IEEE Second International Conference on Multimedia Big Data (BigMM). IEEE, 237–240, 2016.

[35] Susmita Ray, "A Quick Review of Machine Learning Algorithms", 2019 International Conference on Machine Learning, Big data, Cloud and Parallel Computing, pp. 35–39, 14–16 Feb. 2019.

[36] K. Thirunavukkarasu, Ajay S. Singh, Prakhar Rai, Sachin Gupta, "Classification of IRIS Dataset using Classification based KNN Algorithm in Supervised Learning", 2018 4th International Conference on Computing Communication and Automation, 2018, pp. 1–4.

[37] Mahesh Ashok Mahant, VidyullathaPellakuri, "Innovative supervised machine learning techniques for classification of data", Materials today: Proceedings, February 2021.

[38] R. Saravanan, Pothula Sujatha, "A State of Art Techniques on Machine Learning Algorithms: A Perspective of Supervised Learning Approaches in Data Classification", Second International Conference on Intelligent Computing and Control Systems, June 2018.

[39] Z. Yi, H. Zhang, P. Tan, M. Gong, "Dual GAN: Unsupervised Dual Learning for image-to-image translation", Proc. ICCV, pp. 2868–2876, 2017.

[40] Zhiqiang Ge, "Semi-supervised data modelling and analytics in the process industry: current research status and challenges", IFAC Journal of systems and control", Vol. 16, June 2021.

[41] AvgoustinosVouros, Eleni Vasilaki, "A semi-supervised sparse K-means algorithm", Pattern Recognition Letters, Vol. 142, February 2021.

6

Signal Processing for OFDM Spectrum Sensing Approaches in Cognitive Networks

Ishrath Unissa[1], Md. Aleem[2], and Syed Jalal Ahmad[3]

[1]CMR Technical Campus, Secunderabad
[2]Bhaskar Engineering College, Moinabad
[3]Malla Reddy Engineering College (main campus), Secunderabad
E-mail: ishrathunnisa94@gmail.com; Mdaleem80@gmail.com;
jalalkashmire@gmail.com

Abstract

To sense the spectrum in cognitive networks is an important aspect due to various complications occurs between end users, particularly when the transmission is orthogonal frequency division multiplexing (OFDM). The cognitive radio (CR) technology is the most effective approach used to observe the confusion between spectrum utilization and spectrum scarcity. The fundamental approach to sense the spectrum is energy detection approach in OFDM technology due to its simplicity and has been widely used in CR. However, this approach is very much uncertain in the presence of noise. To avoid this drawback, another approach of OFDM was tested called "cyclostationary" to sense the spectrum, but this type of OFDM approach increases computational complexity for real-time applications. This chapter presents the enhanced and comparative analysis of OFDM spectrum sensing approaches in cognitive networks.

Keywords: spectrum sensing, cognitive radio networks, OFDM, primary and secondary users.

6.1 Introduction

Cognitive radio networks (CRN) is a type of radio network that can dynamically utilize the spectrum channels efficiently as the users are increasing on regular basis and the spectrum is limited. It can effectively utilize available (ideal) channels in its range for the transmission and reception of signals and change its parameters accordingly. The CRNs consists of two users; one is primary user (PU) or licensed user, who has license to access the channel and another is secondary user (SU) or unlicensed user, who access the channel without any license when the channel is ideal or not in use by the PU. To use the channels, the CRNs needs to perform a cycle known as cognitive radio cycle [1]. This cycle consists of four stages namely

   i. Spectrum sensing
      In this stage, the less-utilized channel is detected, that is whether the PU is using the channel or not.
  ii. Spectrum management
      Whether the detected channel is appropriate to communication is decided in this stage.
 iii. Spectrum sharing
      If the channel is appropriate, then it is shared by the SU for the transfer of the data.
  iv. Spectrum mobility
      In this stage, the SU need to migrate from one channel to another ideal channel if the primary channel on which it is operating gets active or wants to use the channel.

All these stages have their own importance in the CRNs. In this chapter, the spectrum sensing is discussed as it is the crucial and the first step in the cycle. This stage should provide a greater number of channels to the SU to avoid the interference with the PU s. Hence, the spectrum must be monitored continuously for this purpose. However, there are some issues in sensing the spectrum as discussed in [2, 3] namely the uncertainty in noise, the uncertainty in channel, sensing interference limit, and aggregate interference uncertainty.

   i. Uncertainty in noise: Power of the noise should be known for the calculation of sensitivity which is practically impossible. This problem can be solved by taking into account the worst case noise hypothesis to design the detector which is more sensitive. The sensitivity can be represented as [3].

$$\text{Sensitivity} = \frac{P_T L_P (I_R + M_D)}{N_P}, \tag{6.1}$$

where P_T is the power transmitted, L_P is the loss of path, I_R is the range of interference, M_D is the maximum distance between the transmitter and receiver of the PU, and P_T is the power of noise. Hence, Equation (6.1) also represents the minimum signal to noise ratio (SNR).

ii. Uncertainty in channel: There may be uncertainty in the strength of the received signal from which the SU may get false perception that the PU is active. The SU should take noise into the consideration while sensing the spectrum. However, there is a tradeoff between the avoidance of interference and efficiency of the spectrum. If the spectrum observation time is increased, then the efficiency of spectrum is decreased and if the efficiency of the spectrum is increased by taking more time for transmission, there is more interference [4].

iii. Sensing interference limit: The interference is caused by the PU when it gets active and the SU is communicating using its channel. The reasons for this kind of situation are (a) the position of the receiver of the PU is not known by the SU and (b) the SU may not have the information about the inactiveness of the receiver. Such information plays important role in sensing the spectrum.

iv. Aggregate interference uncertainty: If the number of the SUs increases, the overall interference in detecting the ideal spectrum also increases as they may act simultaneously on the primary spectrum. Therefore, the spectrum sensing should be more accurate.

6.1.1 Spectrum Sensing in CRNs

There are several methods used for spectrum sensing in CRNs. The most commonly used spectrum sensing methods include energy detection method, matched filter detection method, cyclostationary detection method, and optimized spectrum sensing methods. The energy detection method [5] is noncooperative and PU independent method that is, it does not require any information of the PU parameters such as shape of the signal, modulation technique used, and format of the packets. This method is widely used due to its simplicity. However, this approach is more susceptible to uncertainty in noise. In matched filter method [6], the information about PU parameters is required and the maximum gain is achieved in less time as it considers the correlation of unknown signals with the known signals. The modulated signals can be separated from the nose signals at small SNR values in cyclostationary detection [7]. But this method requires PU information partially. Optimized spectrum sensing methods [8] combine more than one general method to optimize the parameters obtain from the spectrum information accurately. These methods require regular information update and have computational complexities.

A number of methods were discussed by the research community to detect the ideal spectrum for the communication of SUs. Pandit and Singh [9] surveyed about the spectrum sensing methods for efficiently utilizing the spectrum and the issues incorporated with spectrum sensing. Further, they presented a hybrid noncooperative method for spectrum sensing as well as a cooperative spectrum sensing method along with their advantages and disadvantages. The application of a particular method is based on the parameters chosen and the PU. Kang [10] discussed about the challenges in spectrum sensing, different spectrum sensing techniques, requirements, and considerations in spectrum-sensing stage. The tradeoff between the parameters is also represented to obtain the minimal value of parameters, such as sensing complexity and time for efficient spectrum sensing. The detection method proposed by Ashish and Soo [11] consists of two detectors; one is energy detector and second is preamble detector and two threshold levels are set. First the energy detector is used to make the local decision and if the value obtained by it lies in between the threshold limits then the preamble detector is activated to ensure the presence or absence of the PU. However, the method used is cooperative as it requires the information of PU parameter.

In this chapter, the signal processing using OFDM techniques are used for spectrum sensing are discussed.

The rest of the chapter is organized as follows: Section 6.2 presents the literature survey. In Section 6.3, the enhanced OFDM techniques for spectrum sensing in CRNs and their comparative analysis is discussed and the chapter is concluded in Section 6.4.

6.1.2 Multiple Input Multiple Output OFDM Cognitive Radio Network Technique (MIMO–OFDMCRN)

This technique of spectrum sensing is proposed by Shan Jin and Zhang [12]. In this approach, the authors use PU of OFDM in MIMO environment by considering "P_N" transmitting aerials and "f_N" subcarriers for transmission in every MIMO aerials. The signal model used in this approach is called sparse model. So, for every antenna, the signal type is sparse with limited subcarriers.

Let the subcarrier number be "f_N" and "S_N" be the antennas required to receive the signal for SU in MIMO–OFDM system. To use the signal of spectrum holes efficiently, it is very necessary for the SU to sense the PUs on priority. This is done by measuring the transmitting signal of the PU at the receiver. The subcarrier frequencies can be calculated by the SU directly. In this approach, the received signals are initially modulated by the application of random signal processor, and then mixed with the incoming signal. The resulting signals are sampled using analog to digital converter (ADC). To recover the actual signal coming from the PU, a digital signal processor is used at the SU. To achieve this goal an algorithm is designed to recover the actual signal. The designed algorithm here is called running recovery algorithm. The algorithm has the ability to use directly the frequency domain.

A. Channel and Transmission Model

This part of the proposed approach presents the description of the channel and transmission model.

Let "$B_{p(n)}$" be the "n^{th}" OFDM transmitted signal coming from the "p^{th}" aerial. Here, the authors considered "B_p" as OFDM modulated symbol, which has been given to every subcarrier, latter then symbols are organized by inverse fast Fourier transform (IFFT) technique. The "i^{th}" coefficient of the above technique at the "j^{th}" aerial can be represented as

$$X_j(i) = \frac{1}{\sqrt{f_N}} \sum_{n=1}^{f_N} B_p(n) e^{\frac{j2\pi n(n-1)}{f_N}}, \qquad (6.2)$$

where "$B_{p(n)}$" represents the "n^{th}" subcarrier modulated signal of the "n^{th}" aerial and "$j = \sqrt{-1}$." In the proposed approach, whenever the PU considers the OFDM–MIMO channel, few subcarriers are allotted to PU to work smoothly. However, here we consider only "f_N" elements of nonzero are used in "B_p". So signal transmission of the "n^{th}" aerial can be represented as

$$X_j = IF_{f_N}^{-1} B_P. \qquad (6.3)$$

Here, "IF_{f_N}" represents the matrix of IFFT. Now "X_j" signal is given to the receiver through wireless channel for transmission. In this approach, the impulse channel response is set in between "n^{th}" transmitting aerial and "j^{th}" receiving antenna. Here, we choose "k" number of multiple paths including wireless channels in between the transmitting and the receiving aerials "$a_{i,j}$"

can be represented as

$$a_{i,j} = \begin{bmatrix} a_{i-j} \\ \vdots \\ a_{i-j}(k-1) \end{bmatrix}. \tag{6.4}$$

So, the received signal can be represented as

$$Y_{i-j} = a_{i-j}X_j. \tag{6.5}$$

Here, "Y_{i-j}" representing transmitting signals coming from "n^{th}" aerial and taken by the "j^{th}" aerial for reception. "A_{i-j}" represents the result of the cyclic convolution in matrix form, and can be represented as

$$\begin{pmatrix} a_{i,j}^{(0)} & 0 & \cdots & 0 & a_{i,j}^{(k-1)} & \cdots & a_{i,j}^{(1)} \\ \vdots & a_{i-j}^{(0)} & \ddots & \vdots & 0 & \ddots & \vdots \\ a_{i,j}^{(k-1)} & \vdots & \ddots & 0 & \vdots & \ddots & a_{i,j}^{(k-1)} \\ 0 & a_{i,j}^{(k-1)} & \cdots & a_{i,j}^{(0)} & 0 & \vdots & 0 \\ \vdots & 0 & \ddots & \vdots & a_{i,j}^{(0)} & 0 & \vdots \\ \vdots & \vdots & \ddots & \ddots & \vdots & \ddots & 0 \\ 0 & 0 & \cdots & 0 & a_{i,j}^{(k-1)} & \cdots & a_{i,j}^{(0)} \end{pmatrix}. \tag{6.6}$$

As in this case, we consider "P_N" and "S_N" transmitting and receiving aerials, respectively, so all the "P_N" transmitting signals are received by "j^{th}" aerial and can be represented as

$$Y_j = \sum_{i=1}^{PN} Y_{i,j}, j = 1, \ldots S_n. \tag{6.7}$$

So, the signal received coming from entire set of aerials can be represented as a matrix-vector and can be given as

$$\begin{bmatrix} Y_1 \\ Y_2 \\ \vdots \\ Y_{SN} \end{bmatrix} = \begin{bmatrix} A_{1,1} & \cdots & A_{PN,1} \\ A_{1,2} & \cdots & A_{PN,2} \\ \vdots & \cdots & \vdots \\ A1_{SN} & \cdots & A_{PN,SN} \end{bmatrix} = \begin{bmatrix} X_1 \\ X_2 \\ \vdots \\ X_{SN} \end{bmatrix} + U. \tag{6.8}$$

Here, "U" is a noise. For simplicity Equation (6.8) can be represented as

$$Y = AX + U, \tag{6.9}$$

where "A" is a matrix used to represent channel of Equation (6.8)

B. Mathematical Model of Sparse Signal

As in the proposed approach, we consider number of aerials in the receiver side, so multiple number of ADCs is used. As a result of which sparse model is used for OFDM–MIMO system using CR technique. In this technique, a lone ADC is used to sample the signals coming from multiple receivers by mixed approach, and can be represented as

$$Y_R = \sum_{j=1}^{S_N} D_j Y_j^R. \tag{6.10}$$

Here, "Y_j^R" is the received signal of "j^{th}" aerial without sampled by ADC and "D_j" represents the vector of the modulated received signal of every individual antenna. Also, in this approach, we use pseudo-random approach to modulate the signal, as this type of modulation technique is simple, easy to compatible with the hardware. For easy operation the random binary value is chosen as "± 1" from "$D_{j(n)}$." A very common technique of signal generation is used here called Gold Sequence to generate random sequence. Then Equation (6.10) can be re arranged as

$$Y_R = \begin{bmatrix} d_1 d_2 & \cdots & S_N \end{bmatrix} \begin{bmatrix} Y_1^R \\ Y_2^R \\ \vdots \\ Y_{SN}^R \end{bmatrix}, \tag{6.11}$$

where "D_j" is matrix diagonal and can be represented as

$$D_j = \begin{bmatrix} D_j(1) & 0 & \cdots & 0 \\ 0 & D_j(2) & \vdots & \vdots \\ \vdots & \vdots & \ddots & 0 \\ 0 & \cdots & 0 & D_j(f_N) \end{bmatrix} \tag{6.12}$$

Equation (6.12) can also be expressed as

$$D = \begin{bmatrix} D_1 & D_2 & \cdots & D_{SN} \end{bmatrix} \tag{6.13}$$

By the use of Equations (6.3), (6.8), and (6.9), we have

$$
\begin{bmatrix} Y_1^R \\ Y_2^R \\ \vdots \\ Y_{SN}^R \end{bmatrix} = A \begin{bmatrix} IF_{F_N}^{-1} & 0 & \cdots & 0 \\ 0 & IF_{F_N}^{-1} & \vdots & \vdots \\ \vdots & \vdots & \ddots & 0 \\ 0 & \cdots & 0 & IF_{F_N}^{-1} \end{bmatrix} \begin{bmatrix} B_1 \\ B_2 \\ \vdots \\ B_{SN} \end{bmatrix} + U. \qquad (6.14)
$$

So, the sampled and received process can be represented as

$$
YR = DA \begin{bmatrix} IF_{F_N}^{-1} & 0 & \cdots & 0 \\ 0 & IF_{F_N}^{-1} & \vdots & \vdots \\ \vdots & \vdots & \ddots & 0 \\ 0 & \cdots & 0 & IF_{F_N}^{-1} \end{bmatrix} \begin{bmatrix} B_1 \\ B_2 \\ \vdots \\ B_{SN} \end{bmatrix} + V
$$

$$
DAIF - 1B + v = CB + v, \qquad (6.15)
$$

where "v" = noise equivalent, "B_i" = transmitted symbol representation of OFDM for "i^{th}" aerial. Due to limited usage of subcarriers "B_i" is only sparse for a limited nonzero number of rudiments on the vector and can be defined as

$$
B = DAIF^{-1}. \qquad (6.16)
$$

Also, to sense the spectrum, the existing mathematical approach can be represented as

$$
Min|B|_{L1}] \quad S.T.Y_R = CB + v. \qquad (6.17)
$$

However, here we use "L_P" instead of "L_1" to get optimum performance. So, the diversity of "L_P" can be represented as

$$
J(P, B) \approx sgn(P) \sum_{j=1}^{P_N f_N} |B(j)|^P \qquad (6.18)
$$

Here, P = real value number
$P \in |0, 2|$ and can be represented as

$$
B^* = arg_B \min H(B), \qquad (6.19)
$$

where

$$
H(B) = \|\psi_B - Y_R\|^2 L_2 + \phi j(P, B), \quad \text{for } \phi = \frac{2\omega}{|P|} \qquad (6.20)
$$

Algorithm to detect spectrum

Input B, Y_R

Initialize $\omega = \eta, j\varepsilon[1, \ldots, SN], n = 1$
Repeat

$Z_n = \text{diag}(B_n - 1(j)^{1-P/2})$

$E_n = Z_n(B_n^T B_n + \omega I)^{-1} B_n Y_R$

Where $B_n = BZ_n$ with $\omega \geq 0$

$B_n = Z_n E_n$

Until $\|B_n - B_{n-1}\|_2 < \theta$

Compute $\Omega_B = \arg\max(|B_n|, S, B^* = B_n$

S = number of zero elements in B_n

Output Ω_B, B^*

In the algorithm, "n" represents the iterations, "B^*" is the signal estimation. Here, value of "θ" is chosen as "0.005."

Observations

In the proposed approach, for analysis purposes, two transmitting and receiving aerials have been used to validate the experimental results. From these results, the following observations have been carried out

1. The probability of detection is more when the subcarriers occupied are less.
2. The probability of the signal recovery increases when SNR value is high.
3. When SNR is greater than "8dB" and subcarrier occupied is "1" then the detection probability is more than "98%."
4. The relative mean square error is minimum when subcarriers occupied are less. Thus, this error increases when active subcarriers are more.
5. The lesser the number of active subcarriers, the lesser will be the error in the recovery of data.
6. The bit error rate is almost constant even if the SNR is varied when occupied subcarriers are least.
7. The probability of detection is higher compared to art of work, when lesser number of subcarriers is occupied.

6.1.3 Improved Sensing of Cognitive Radio for MB Spectrum using Wavelet Filtering

In this approach, an improved spectrum sensing method has been proposed using wavelet filtering technique for CR [13]. Initially, the proposed method detects the holes of the spectrum by measuring the value of the power spectrum density (PSD) of the recovered signal. This method also calculates the contours of the frequency allotted sub-bands of a wide spectrum band. The variation in the PSD finds the boundaries of the allotted frequencies. So, to detect these variations of the frequencies, wavelet transform (WT) approach is used to calculate the PSD by convoluting the initial derivative of the wavelet and the PSD to get the maxima. As a result of which reduces the complexity of the system and improves the speed of the system performance. The explanation of the proposed approach is presented in following section.

6.1.3.1 MB-Spectrum Sensing Method

Figure 6.1 presents the summary of the presented approach that is improved spectrum sensing technique.

Fundamentally, this approach finds the PSD of the allotted band and estimates the contours of the spectrum by the use of Gaussian flittering. Then it generates a cover for the spectrum of the allotted sub-bands. The explanation here includes (6.1.3.1.1)-estimation of PSD; (6.1.3.1.2 and 6.1.3.1.3)-edge detection (a) and (b); (6.1.3.1.4)-edge classifier; (6.1.3.1.5)-corrections of errors; and (6.1.3.1.6)-generation of spectrum mask.

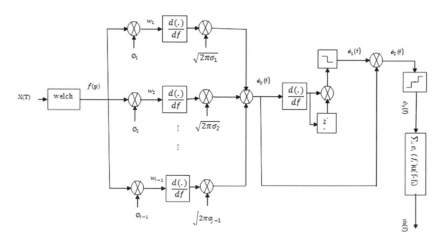

Figure 6.1   Block of MBSS.

6.1.3.1.1 Estimation of PSD

Spectrum sensor initially estimates the PSD, here, the Walch's method [14] is used to get the estimation of the PSD with number of overlapping of a signal segment to get the optimum result of spectrum estimation. Initially, "M" number of samples has been taken at a rate of "S_f" samples. These samples are bifurcated into "Q" levels having length "l" with "C" samples overlapped.

Few practical aspects are very necessary to consider, getting the optimum results for the system performance. Firstly, the range of frequency must be bounded with "S_f" samples. The accuracy of the grid frequency is proportional to the total number of steps of the PSD, which affects the spectrum in terms of the precision location of the spectrum.

6.1.3.1.2 Edge detection (a)

The estimate of PSD is sent through say "j" parallel divisions. Where the PSD estimated signals are filtered and differentiated. The signal at "q^{th}" division can be represented as

$$W_{\bar{q}}(f) = \frac{d}{df}(\theta_q(f) * \widehat{PSD}(f)). \qquad (6.21)$$

It is observed that the original signal function is convolved with the orthogonal family functions with variable scales. Equation (6.21) depicts the positive peaks for rising edges and negative peaks for falling edges. The resultant signals of each division is multiplied (Equation (6.22)) as it restores all the peak values obtained at each division and small peaks in same divisions are dismissed. To preserve the falling edges, "j" is considered to be odd and hence the negative peaks not lost. The Gaussian filters used are blurring with variable scales say "ϕ_q" with the impulse response represented as

$$\bar{E}_1(f) = \prod_{q=0}^{j-1} \sqrt{2\pi\phi_q^2 W_{\bar{q}}(f)} \qquad (6.22)$$

$$\theta_q(t) = \frac{1}{\sqrt{2\pi\phi_q^2}} \exp\left(\frac{-f^2}{2\phi_q^2}\right), \qquad (6.23)$$

where $q = 0, 2, \cdots j–1$.

The scales "ϕ_q" plays vital role in detection of edges and hence its value can be represented as

$$\phi_q = 100.\frac{S_f}{N}2^{-q}. \qquad (6.24)$$

The effective length of the filter is bounded according to the scale and implies the exact PSD estimated limit by "s_f" and "N" samples.

6.1.3.1.3 Edge detection (b)

To find the precise raising or falling edge location of the spectrum sensed, the resultant of the edge detection (a) is differentiated again to obtain the extremes locally of Equation (6.23) represented as

$$\bar{E}_2(f) = \begin{cases} s(f - f_q), & \bar{E}_1(f)\bar{E}_1(f - S_{\frac{f}{N}}) \\ 0, & \text{otherwise} \end{cases} \tag{6.25}$$

This bisection method of finding the roots is applied to extract the location of edges. The two points of "E1" which are adjacent are multiplied, if the output is negative, then there is a sign transition and root is present at these points, else there is no root. It is still significant to detect whether the edge is falling or rising or mixed after locating them. The function represented in Equation (6.26) gives the information of the edges "f_q" are the edges occurring frequencies.

$$\bar{E}_3(f) = \bar{E}_1(f)\bar{E}_2(f) = \bar{E}_1(f_q) \sum_q s(f - f_q) . \tag{6.26}$$

6.1.3.1.4 Edge classifier

As the signals are affected by noise, false-edge detection is possible and it should be ruled out frequently. For this purpose, the estimated signals are passed through the decision with the set threshold value as "ϵ." Hence, the impulses of the edge can be represented as

$$\bar{E}_H(f_q) = \begin{cases} \delta(f - f_q), & \bar{E}_3(f_q) > \epsilon \\ \delta(f - f_q), & \bar{E}_3(f_q) > -\epsilon \\ 0, & \text{otherwise} \end{cases} \tag{6.27}$$

This equation can be used to generate the mask of bands (spectrum).

6.1.3.1.5 Correction of errors

Miss-edge detection still persists even after the hard decision; however, these can be reduced by considering the fact that no two consecutive edges have the sign same. If large raising and falling edges are detected then the edges present at the left and right must be eliminated, respectively. If the entire band of PUs falls in the range of the spectrum sensing, then even number of edges should be present. Hence, the performance of the detector can be improved.

6.1.3.1.6 Generation of spectral mask

The spectral mask with two levels is obtained with high and low levels which represents the occupied and ideal sub-bands, respectively. The edges are modeled as unit impulses. The spectrum mask formed can be represented as

$$\bar{N}(f) = \int_0^f \bar{E}_H(f_q)df = \sum_q^0 \bar{E}_H(f_q)u(f - f_q). \qquad (6.28)$$

6.1.3.1.7 Sensing of OFDM signals

In the previous section, the spectrum-sensing method discussed can be applied to the OFDM signals, as these signals are renowned in digital communication. Let the rectangular subcarriers present I OFDM be "SC_n" with bandwidth of "BW_n" then the PSD of the complex mask (cover) can be represented in [15] as

$$PSD(f) = B \sum_{j=0}^{SC_n-1} \sin c^2 \left(\frac{f - f_z}{BW_n/2} \right), \qquad (6.29)$$

where "f_z" is the position of subcarriers calculated as

$$f_z = \frac{BW_n}{2} \left(\frac{j - SC_n - 1}{2} \right). \qquad (6.30)$$

Although the Equation (6.29) represents the infinite OFDM PSD signal but few OFDM symbols can be taken into account and the filtered version of this signal is complex and difficult to evaluate. Hence, the approximate of $\sin c^2(.)$ function by Gaussian as

$$PSD(f) \cong \sum_{j=0}^{SC_n-1} \exp \left(\frac{f - f_z}{2\beta^2} \right). \qquad (6.31)$$

These approximations are improved when the subcarriers number increases.

From above discussion two observations can be made which are discussed as

(a) Signal peak occurs on the edges of transition bands with '$\pm B^j$' as peak amplitude and the limit of threshold is "$0 \leq \epsilon \leq B^j$." The value of "B" is not known and is hardly unit, then the unit of threshold id "$0 \leq \epsilon \leq 1$."

(b) The average of "ϕ_q^2" is considered as harmonic mean of each scale of filter division and if the values of "ϕ_q^2" obtained are varied from each other, then the mean is minimum value of "ϕ_q^2."

Observations:

1. The PSD estimation is precisely evaluated in the presence of noise.
2. The probability of spectrum detection is optimum when compared to the existing works.
3. This method evaluates the contour of the maxima.

6.1.4 OFDM-Based Blind Sensing of Spectrum in Cognitive Networks

Simin et al. [16] proposed a blind method for spectrum sensing in CRN using OFDM technique. In this approach, a constrained generalized likelihood ratio test (CGLRT) which is multipath based and is used to improve the spectrum sensing in CRN. The proposed approach performs better than the cyclic prefix correlation coefficient (CPCC) which is used to sense the spectrum in multipath ambiance. The authors develop a hybrid method combining the above mentioned algorithms for better results and they also discussed an efficient GLRT algorithm for unsynchronized OFDM signals which gives approximately same results as synchronized detected OFDM signals employed in multi route ambiance.

6.1.4.1 Model of the Proposed System

The model proposed in this approach is similar to the method proposed in [17, 18]. The number of subcarriers employed in the primary system of OFDM are assumed to be "N." The synchronization of PUs and cognitive radios are perfectly maintained. In the 1th OFDM part the complex symbol to be transmitted is $\{K_{1,s}\}_{s=0}^{n-1}$ with its expectation as "ϕ_K^2." Hence, the modulated signal can be represented as

$$m_1(i) = \frac{1}{\sqrt{N}} \sum_{s=0}^{N-1} K_{1,s} e^{\frac{j2\pi is}{N}}, \tag{6.32}$$

where "1 = 0, 1, ... N-1." When large number of "N" is considered the Equation (6.32) will be equal to the complex random Gaussian variable with zero mean circular symmetry. Whose variance is equal to that of "ϕ_K^2."

The transmitted signal of the 1th OFDM part along with the guard signals can be represented in vector form as

$$m_1 = [m_1(N-1)\ldots m_1(0)m_1(N-1)\ldots m_1(N-N_j)]. \qquad (6.33)$$

The term "$m_1(N-1)\ldots m_1(N-N_j)$" is cyclic prefix. Hence, the noise and received vectors of their signal can be given as

$$n_1 = [n_1(N-1)n_1(N-2)\ldots n_1(0)n_1(-1)\ldots n_1(-N_j)]^{\text{T}} \qquad (6.34)$$

$$r_1 = [r_1(N-1)r_1(N-2)\ldots r_1(0)r_1(-1)\ldots r_1(-N_j)]^{\text{T}}, \qquad (6.35)$$

where the samples of noise are independent and identically distributed (i.i.d) complex Gaussian random variable.

The signals from the PUs are received by multipath faded channel wirelessly having the filter taps of the channel as "h_q," where "q" $=1\ldots N_t$ and N_t is defined as the count of components present in the multipath. The assumption is made that the fading remains unchanged in the and sensing interval. Thus, the relation between the transmitted, noise and the received signal can be represented as

$$r_1 = h\, h_{\overline{ml}} + n_1, \qquad (6.36)$$

where $\overline{ml} = [m_l^{\text{T}}, m_{l-1}(N-1), \ldots m_{l-1}(N-N+1)]^{\text{T}}$ and the matrix of toeplitz for the filter channel can be represented as

$$h = \begin{pmatrix} h_1 & \ldots & h_{Nt} & 0 & \ldots & \ldots & \ldots & 0 \\ 0 & h_1 & \ldots & h_{Nt} & 0 & \ldots & \ldots & 0 \\ \vdots & & \ddots & & \ddots & & & \vdots \\ \vdots & \ldots & 0 & h_1 & \ldots & h_{Nt} & 0 & 0 \\ \vdots & & \ldots & \ddots & \ddots & & \ddots & 0 \\ 0 & \ldots & \ldots & \ldots & 0 & h_1 & \ldots & h_{Nt} \end{pmatrix} \qquad (6.37)$$

The inter symbol interference block of "r_1" are the last samples of "N_j." Therefore, based on the Equation (6.36), $\phi_r^2 = \phi_k^2 \sum_{q=1}^{N_t} |h\, h_q|^2 + \phi_n^2$ is the variance of "$r_1(p)$" when the PU is active and when the PUs are not active then "$\phi_r^2 = \phi_n^2$" and when the PUs are active the SNR can be represented as

$$\text{SNR} = \phi_K^2 \sum_{q=1}^{N_t} |h\, h_q|^2 \Big/ \phi_n^2.$$

The hypothesis used for detection of ideal and active state of PU are "I_0 and I_1," respectively

$$\begin{cases} I_0 & \text{if} x \leq \tau \\ I_1 & \text{if} x > \tau \end{cases} \tag{6.38}$$

where "τ" is the threshold value and "x" is the statistics tested. The two probabilities that is the probability of detecting that the PU is active and is expressed as "D_p" and the probability of false alarm that is the probability of detecting PU as active, when it is in ideal state is expressed as "f_p" and these probabilities can be represented as

$$D_P = P\{X > \tau/I_0 \tag{6.39}$$

$$f_P = P\{X > \tau/I_1. \tag{6.40}$$

The generalized likely hood ratio test for sensing of spectrum is mentioned in [19]. This test implied in different parametric assumptions, such as noise variance and covariance signal matrix which are unknown. The test is considered in its generalized form and is applied to detect the OFDM signals.

Consider the "A^{th}" length column vector with the covariance matrix "B_a" and noise variance as "ϕ_C^2" of the "I^{th}" part of the received signal where both the parameters are unknown then GLRT can be represented as [16].

$$\text{LRT}(a) = \frac{\frac{f_a}{I_1}, \overline{B_a}\left(\frac{a}{I_1}, \overline{B_a}\right) > I_1}{\frac{f_y}{I_0}, \phi_C^2\left(\frac{a}{I_0}, \phi_C^2\right) > I_0} \ \varepsilon, \tag{6.41}$$

where "$\overline{B_a}$" and "ϕ_C^2" are the highest likelihood estimates and can be represented as

$$\overline{B_a} = \max_{B_a \in k_{B_a}} \text{In} \frac{f_a}{I_1}, \ B_a\left(\frac{a}{I_1}, B_a\right) \tag{6.42}$$

and

$$\overline{\phi_C^2} = \max_{\phi_C^2} \left\{ \text{In} \frac{f_a}{I_0}, \ \phi_C^2\left(\frac{a}{I_0}, \phi_C^2\right) \right. \tag{6.43}$$

After substituting the Equations (6.42) and (6.43) in (6.41), we get the GLRT as

$$\begin{aligned} X_{\text{LRT}}(a) &= \{\text{LRT}(a)\}^{\frac{1}{LA}} \\ &= \frac{\frac{1}{A}\text{tr}(\overline{B_a})}{\det^{\frac{1}{A}}(\overline{B_a})} \eta(\overline{B_a}), \end{aligned} \tag{6.44}$$

where

$$\eta(\overline{B_a}) = \frac{\exp(-1/A \operatorname{tr}(\overline{B_a} - B_a))}{\exp(-1)} \tag{6.45}$$

$\overline{B_a} = \frac{1}{L_{aa}^H}$ is the covariance matrix Hermitian transpose.

The variations in "B_a" under meek conditions will have the same structure as "B_a" satisfying $\operatorname{tr}(\overline{B_a}^{-1} B_a) = A$, which implies that $\eta(\overline{B_a}) = 1$ and the Equation (6.44) becomes

$$X_{LRT}(a) = \frac{\frac{1}{LA} \operatorname{tr}(aa^H)}{\det^{\frac{1}{A}}(\overline{B_a})} > \frac{I_1}{I_0} \tau. \tag{6.46}$$

If unconstrained covariance matrix structure is considered then Equation (6.46) can be modified as

$$X_{LRT}(a) = \frac{\frac{1}{A} \operatorname{tr}(\overline{B_a})}{\det^{\frac{1}{A}}(\overline{B_a})} > \frac{I_1}{I_0} \tau. \tag{6.47}$$

From (6.47), we can conclude that by focusing on specified part of the observations, the properties of covariance matrix structure can be used to improvise the GLRT assessment. The performance of cyclic prefix correlation is not much effective in multipath propagation channels as shown in [16], so for the OFDM signals received can be estimated by excluding the inter symbol interference (ISI) part as in cyclic prefix the correlation can be obtained at the top and end of each received OFDM block. Therefore, the estimation of covariance matrix is improvised by using GLRT based on the correlation of multipath.

6.1.4.2 Constrained GLRT Algorithm

In this algorithm the OFDM block received excluding the ISI with the transmitted and noise signals can be represented as

$$r_{l1} = [r_1(N-1), \ldots, r_1(0)]^T$$

$$m_{l1} = [m_1(N-1), \ldots, m_1(0)]^T$$

and

$$n_{l1} = [n_1(N-1), \ldots, n_1(0)]^T.$$

And can be related as

$$r_{l1} - h1m_{l1} + n_{l1},$$

where "h1" is the circular channel matrix similar to that of Equation (6.38), the covariance matrix can be represented as

$$R_{r1} = h_1 h_1^H \phi_k^2 + \phi_n^2 I_d, \tag{6.48}$$

where "I_d" denotes the identity matrix

The properties of 'R_{r1}' are as follows

I. It is Hermitian.
II. Its diagonal elements are same.
III. It has the property circular similarity.

The GLRT can be calculated using Equation (6.46) and substituting L = N

$$X_{LRT}(r_1) = \frac{\frac{1}{LA} \mathrm{tr}(r_1 r_1^H) > I_1}{\det^{\frac{1}{A}}(\bar{R}_{r_1}) < I_0} \tau_1. \tag{6.49}$$

The correlation values obtained has maximum delay of "$N_j - 1$" samples.

6.1.4.3 A Multipath Correlation Coefficient Test

It can also be used instead of primary signal correlation property to detect the noise in the back ground as shown in [16, 20–22]. The combination of CPCC and MPCC will give better information about the presence of the PU. However, care should be taken by considering the proper threshold value for overall false alarm.

6.1.4.4 Probability Calculation

The previous algorithms discussed are timely synchronized; however, if there is no synchronization of symbols. The cyclic correlation can be evaluated by taking into account the symbols placed in the delays of $\pm N$ [18]. To efficiently calculate the GLRT for a synchronized OFDM signals the last part of the samples, that is "$N1 = N - N_j + 1$" in the block of each "N" samples is considered. From the previous discussion it has been observed that it is not necessary that the signals are aligned with the OFDM transmitted blocks due to the reason that the signals are not synchronized. In this model every sample in the vicinity of the transmitted vectors are closely related with the mean correlation.

Observations

I. The three algorithms proposed namely CPCC, MPCC, and their combination is compared with the performance of the energy detector.

II. The performance of MPCC is approximately similar to that of the combined algorithm and is better than CP for channel taps "$N_t = 8$."

III. The system performance of GLRT, CPCC, and MPCC is better when compared with the art of work.

IV. At different values of "N_t and N_j" the detection performance for both MPCC and CPCC varies minutely. If the value of "N_t" is large then the correlation is better. However, it requires additional resources.

V. If $N_t = N_j = 8$ the combined algorithm that is MPCC and CPCC based on GLRT is having optimum value of the covariance structure when compared to each of a single method separately.

6.1.5 Comparative Analysis

Table 6.1 Comparative analysis

S.No	Techniques			
		Compressive MIMO-OFDM Spectrum Sensing	Wavelet Spectrum filtering in MB	Blind OFDM Spectrum Sensing
1	Approach/method	Received signals are Mixed and sampled by single ADC	Wavelet filtering method	Cyclic prefix and multi-path correlation coefficients based GLRT is used.
2	Detector used	sample detector	Wavelet edge detector	MP-CPCC detectors
3	Sensing technique	Not Cooperative	Not cooperative	Not cooperative
4	Advantages	Efficiently detection of signals. Energy consumption is less, hardware is reduced and in-turn cost is reduced and applied for	Reliable efficient and At low SNRs, maximum performance is achieved	Efficient detection in noisy and multi-path fading environments. Hybrid method perform better when large subcarriers are occupied at optimal SNR.

(Continued)

Table 6.2 (*Continued*)

S.No		Techniques		
		Compressive MIMO-OFDM Spectrum Sensing	Wavelet Spectrum filtering in MB	Blind OFDM Spectrum Sensing
5	Disadvantages	As MIMO systems are used, delays may be present which can disrupt the system performance.	Evaluation of boundaries are critical to calculate the maxima which is the main criteria of the approach	Calculation complexity and threshold must be set appropriately in the hybrid model

6.2 Conclusion

In this chapter, different OFDM spectrum sensing techniques in cognitive radio networks are presented. As spectrum sensing is the vital step in cognitive radio networks the OFDM signals are more efficient to obtain the knowledge of PU's presence and absence in digital communication systems. The methods discussed in this chapter are improved, fundamental and give almost all the information about the existing OFDM spectrum sensing methods. These methods have their own advantages, so one can employ any of these techniques based on their requirements.

References

[1] J. Marinho and E. Monteiro, Cognitive Radio: Survey on Communication Protocols, Spectrum Decision Issues, and Future Research Directions, Wireless Networks, Springer, vol. 18, no. 2, pp. 147–164, 2012.

[2] A. Ghasemi, and E. S. Sousa, Spectrum Sensing in Cognitive Radio Networks: Requirements, Challenges and Design Trade-offs, IEEE Communication Magazine, vol. 46, no. 4, pp. 32–39, 2008.

[3] D. B. Rawat, and G. Yan, Signal Processing Techniques for Spectrum Sensing in Cognitive Radio Systems: Challenges and Perspectives', First Asian Himalayas International Conference on Internet - The Next Generation of Mobile, Wireless and Optical Communications Networks 2009 (AC-ICI-2009), 2009.

[4] E. Hossain, D. Niyato, and Z. Han, Dynamic Spectrum Access and Management in Cognitive Radio Networks, Cambridge University Press, 2009.

[5] K. Ben Letaief and W. Zhang, Cooperative communications for cognitive radio networks. Proc. IEEE. 97(5), 878–893, 2009.

[6] D. Tian, L. Zheng, J. Wang, and L. Zhao, in ConsumerElectronics, Communications and Networks (CECNet), 2012 2nd International Conference On A new spectrum sensing method for OFDM-based cognitive radios (Yichang, China), pp. 812–815, 2012.

[7] M. Derakhshani, T. Le-Ngoc, and M. Nasiri-Kenari, Efficient cooperative cyclostationary spectrum sensing in cognitive radios at low SNR regimes. IEEE Trans. Wireless Commun. 10(11), 3754–3764, 2011.

[8] Unissa I., Ahmad S.J., Radha Krishna P. Optimum Spectrum Sensing Approaches in Cognitive Networks. In: Rehmani M., Dhaou R. (eds) *Cognitive Radio, Mobile Communications and Wireless Networks.* EAI/Springer Innovations in Communication and Computing. Springer, Cham, 2019.

[9] Pandit, Shweta & Singh, Ghanshyam. Spectrum Sensing in Cognitive Radio Networks: Potential Challenges and Future Perspective. 10.1007/978-3-319-53147-2_2, 2017.

[10] B. Kang, "Spectrum sensing issues in cognitive radio networks," 2009 9th International Symposium on Communications and Information Technology, Icheon, pp. 824–828. doi: 10.1109/ISCIT.2009.5341128, 2009.

[11] Ashish Rauniyar and Soo Young Shin. Cooperative spectrum sensing based on adaptive activation of energy and preamble detector for cognitive radio networks. vol. 7, e2, page 1–7, 2018.

[12] Shan Jin and Xi Zhang, 'Compressive Spectrum Sensing for MIMO-OFDM Based Cognitive Radio Network'. IEEE Wireless Communications and Networking Conference (WCNC), pp: 2197–2202, 2015.

[13] Ricardo Tadashi Kobayashi, Aislan Gabriel Hernandes, Mario Lemes Proença Jr. and Tauk Abrao, 'Improved MB Cognitive Radio Spectrum Sensing Using Wavelet Spectrum Filtering Journal of Circuits', Systems, and Computers Vol. 28, No. 8 (2019) 1950136 (22 pages).

[14] P. D. Welch, The use of fast fourier transform for the estimation of power spectra: A method based on time averaging over short, modifed periodograms, IEEE Trans. Audio Electroacoust. AU-15 (1967) 17–20.

[15] T. van Waterschoot, V. Le Nir, J. Duplicy and M. Moonen, Analytical expressions for the power spectral density of CP-OFDM and ZP-OFDM signals, IEEE Signal Process. Lett. 17 (2010) 371–374.

[16] Simin Bokharaiee, Ha H. Nguyen, and Ed Shwedyk 'Blind Spectrum Sensing for OFDM-Based Cognitive Radio Systems'. IEEE Transactions On Vehicular Technology, Vol. 60, No. 3, pp. 858–871, March 2011.

[17] T. Hwang, C. Yang, G. Wu, S. Li, and G. Y. Li, 'OFDM and its wireless applications: A survey,' IEEE Trans. Veh. Technol., vol. 58, no. 4, pp. 1673–1694, May 2009.

[18] S. Chaudhari, V. Koivunen, and H.V. Poor, 'Autocorrelation-baseddecentralized sequential detection of OFDM signals in cognitive radios,' IEEE Trans. Signal Process., vol. 57, no. 7, pp. 2690–2700, Jul. 2009.

[19] T. J. Lim, R. Zhang, Y. C. Liang, and Y. Zeng, 'GLRT-Based Spectrum sensing for cognitive radio', in Proc. IEEE Global Telecommun. Conf., Nov. 2008, pp. 1–5.

[20] W. Zeng and G. Bi, "Exploiting the multi-path diversity and multi-user cooperation to detect OFDM signals for cognitive radio in low SNR with noise uncertainty," in Proc. IEEE Global Telecommun. Conf., Dec. 2009, pp. 1–6.

[21] W. Zeng and G. Bi, "Robust detection of OFDM signals for cognitive UWBinlowSNRwith noise uncertainty," in Proc. IEEE Int. Symp. Pers., Indoor, Mobile Radio Commun., Sep. 2008, pp. 1–5.

[22] E. Cadena Munoz, L. F. Pedraza Martinez and C.A. Hernandez, "Renvi Entropy- Based Spectrum Sensing in Mobile Cognitive Radio Networks using software Defined Radio," Entropy vol. 22, no. 6, pp. 626, 2020.

7

A Machine Learning Algorithm for Biomedical Signal Processing Application

Abhishek Choubey, Shruti Bhargava Choubey, and S.P.V. Subba Rao

Department of Electronics Communication, Sreenidhi Institute of science and technology Hyderabad, India
E-mail: abhishek@sreenidhi.edu.in; shrutibhargava@sreenidhi.edu.in; spvsubbarao@sreenidhi.edu.in

Abstract

In this chapter, significant emphasis has been given to machine learning (ML) for biomedical signal processing. The basic purpose of this chapter is to explore the numerous possibilities of ML in the field of biomedical signal processing. With the increasing volume of available biomedical data, network speed, and computing power, the modern biomedical signal is facing an unique amount of data to interpret and analyze. Phenomena like Big Data-omics restricting from numerous diagnostic methods and novel multiparametric singling modalities tend to produce practically unmanageable amounts of data. Traditional ML ideas have established many limitations when it comes to correctly diagnose like electrocardiogram, hearing aid, and EEG diseases. At the same time, static graph networks are unable to capture the fluctuations in monitor and processing progress. Therefore, deep learning and artificial intelligence are applied in bio-medical singling because they excel at providing quantitative assessments of biomedical singling characteristics.

7.1 Introduction

7.1.1 Introduction to Signal Processing

Computerized signal processing, a field that has its underlying foundations in 17th and 18th century arithmetic, has become a significant device in a large number of assorted fields of science and innovation. The field of computerized signal preparing has filled colossally in the previous decade to incorporate and give firm hypothetical foundations to countless individual territories.

The expression "advanced sign preparing" may have alternate importance for various individuals. For instance, a paired bitstream can be viewed as an "advanced sign" and the different controls or "sign handling," performed at the cycle level by computerized equipment might be understood as "computerized signal preparing." In any case, the perspective taken in this theory is extraordinary. The meaning of computerized signal preparing (DSP) is the idea of a data-bearing sign that has a simple partner. What are controlled are tests of this certainly simple sign. Further, these examples are quantized, which is spoken to utilizing limited exactness, with each word illustrative of the estimation of the example of a (verifiably) simple sign. These controls, or channels, performed on these examples are number-crunching in nature—increments and augmentations. The meaning of DSP incorporates the preparation related to examining, transformation among simple and advanced areas, and changes in word length. Computerized signal processing is worried about the portrayal of signs by arrangements of numbers or images and the handling of these successions. The reason for such preparation might be to assess the trademark boundaries of a sign or to change a sign into a structure that is in some sense more attractive.

Signal handling, when all is said in done, has a rich history, and its significance is apparent in such assorted fields as biomedical designing, acoustics, sonar, radar, seismology, discourse correspondence, information correspondence, atomic science, and numerous others. In numerous applications, as, in electrocardiogram (ECG) examination or frameworks for discourse transmission and discourse acknowledgment, we may wish to eliminate the obstruction, for example, commotion, from the sign or to adjust the sign to introduce it in a structure which is all the more effectively deciphered. As another model, a sign sent over correspondences divert is commonly annoyed in an assortment of ways, including channel contortion, blurring, and the addition of foundation clamor. One of the goals of the recipient is to make up for these unsettling influences. For each situation, handling the sign is required.

Signal preparing issues are not limited to one-dimensional signs. Many picture handling applications require the utilization of two-dimensional sign preparing strategies. This is the situation, for instance, in x-beam improvement, the upgrade and examination of aeronautical photos for identification of timberland flames or yield harm, the investigation of satellite climate photographs, and the improvement of TV transmissions from lunar and profound space tests. Seismic information investigation as needed in oil investigation, tremor estimations, and atomic test observing additionally uses multidimensional sign preparing procedures.

There are numerous reasons why computerized signal handling of a simple sign might be desirable over preparing the sign straightforwardly in the simple area. Initially, an advanced programmable framework permits adaptability in reconfiguring the computerized signal preparing tasks basically by changing the program. Reconfiguration of a simple framework for the most part infers an overhaul of the equipment followed by testing and check to see that it works appropriately.

Exactness contemplations additionally assume a significant part in deciding the type of the sign processor. Resistances in simple circuit parts make it very hard for the framework fashioner to control the exactness of a simple sign handling framework. Then again, a computerized framework gives much better control of precision prerequisites.

Advanced signs are effectively put away on attractive media (tape or circle) without crumbling or loss of sign constancy past that presented in the A/D transformation. As a result, the signs become movable and can be handled disconnected in a distant research center. The computerized signal preparing technique likewise considers the execution of more modern sign handling calculations. It is normally hard to perform the exact numerical procedure on signs in a simple structure, however, these equivalent tasks can be regularly executed on an advanced PC utilizing programming. The advanced execution of the sign preparing framework is quite often less expensive than its simple partner because of the adaptability for alterations. As a result of these focal points, advanced sign preparation has been applied in useful frameworks covering a wide scope of controls.

Be that as it may, advanced execution has its impediments. One viable impediment is the speed of activity of A/D converters and computerized signal processors. Signs having incredibly wide transfer speeds require quick inspecting rate A/D converters and quick computerized signal processors. Subsequently, there are simple signs with enormous data transmissions for which an advanced preparing approach is past the best in the class of computerized equipment.

The methods and utilization of computerized signal handling are extending at an enormous rate. With the coming of huge scope resolution and the subsequent decrease in expense and size of advanced parts, along with speeding up, the class of uses of computerized signal preparing strategies is developing.

7.1.1.1 ECG Signal

The ECG is only the chronicle of the heart's electrical action. The deviations in the ordinary electrical examples show different heart issues. Cardiovascular cells, within the typical state are electrically enraptured. Their inward sides are contrarily charged comparatively with their external sides. These vessel cells will lose their normal pessimism in a very cycle known as depolarization, that is that the major electrical action of the center. This depolarization is unfolded from cell to cell, making an associate inflow of depolarization that may be sent across the total heart. This rush of depolarization delivers a progression of electrical flow and it tends to be known by keeping the terminals on the skin of the body.

Once the depolarization is over, the cardiovascular cells can re-establish their ordinary boundary by a sequencetermed re-polarization. This is furthermore received by the cathodes.

ECG fluctuates as expected, the requirement for a precise depiction of the ECG recurrence substance as per their area in time is basic. This legitimizes the utilization of time recurrence portrayal in quantitative electro cardiology. The quick device accessible for this reason for existing is the Short-Term Fourier Transform (STFT). Yet, the significant downside of this STFT is that its time recurrence exactness is not ideal. Thus, we select a more appropriate method to beat this disadvantage. Among the different time recurrence changes, the wavelet transformation is discovered to be basic and more important. The wavelet change depends on a bunch of dissecting wavelets permitting the deterioration of ECG signal in a bunch of coefficients. Each investigating wavelet has its time length, time area, and recurrence band. The wavelet coefficient coming about because of the wavelet change relates to an estimation of the ECG parts in this time section and recurrence band.

The ECG signal is the time-shifting sign, incorporates the important data identified with heart sicknesses, yet habitually this valuable information is tainted by different clamors. The various commotions liable for debased ECG signals are:

(i) Electromyography noise—emerges from the superposition of strong electrical possibilities over the ECG signal, while wideband commotion is inseparable from the total procurement of ECG blemishes in the electronic hardware. The muscle ancient rarity noise range, bringing about its contortion, covers the range of the QRS complex (5–15 Hz). It is influenced by muscle movement prompted noise.

(ii) Power line obstruction is because of lacking protection from the force supply of ECG securing frameworks and blemishes in the hardware of the speaker. It occurs because of intensity mains' 50–60 Hz pickup and music. Electrical cable mutilation and wideband commotion altogether decline the sign effectiveness, delivering it difficult to recognize fiducial four focuses.

(iii) Baseline ponders baseline meander results from the distinction in terminal impedance because of the patient's happening and pivot, brought about by factor contact and breath among cathode and tissue.

(iv) Electromagnetic radiation—from other electronic gadgets and commotion blended in with other electronic gadgets, regularly at high frequencies.

(v) Motion objects—got from terminal skin contact impedance variety.

(vi) White Gaussian Noise Additive (AWGN)—which has an element of Gaussian likelihood thickness and a component of white force ghostly thickness (noise dissipated over the entire recurrence range) and is *applied sprightly to whatever signal we study.*

It is important and hard to stifle sounds from the ECG signal as commotion defiles the ECG signal. So denoising is the methodology from accessible uproarious information to surmised the obscure sign. The wavelet change gives a portrayal of the sign, decaying it at an alternate goal of time-recurrence. Wavelet change is an appropriate strategy for nonfixed sign handling, for example, ECG.

7.2 Related Work

7.2.1 Signal Processing Based on Traditional Methods

In 2006, once again, Choukari used decomposition at the second level to detect complex QRS, and decomposition at the fourth and fifth levels to detect P and T waves. By measuring MSE and signal-to-noise ratio (SNR), they compared their algorithm's efficiency to db5, db10, coif5, sym6, sym8, and biorth5.5. This algorithm functions efficiently to eradicate different noises at low SNR, but the key drawback is the existence of immense baseline wander. To suppress the baseline phenomenon, a vigorous QRS detection proceduremust be castoff [1].

In the year 2008, in the MATLAB setting, Saritha has identified various forms of irregularities using Daubechies wavelets. To remove ECG characteristics, Ingole used Dyadic wavelet transform, which is stable, exceedingly-competent, precise, and consistent. Afsar suggested a system for categorizing six distinct kinds of strokes from the ECG that is resilient to noise based on

DWT and PCA. Less complexity, strong precision, and time-saving are the merits of this algorithm [2].

In 2008, in the presence of various kinds of noises, Rizzi suggested and introduced an algorithm named R-point detector-based variation of wild parallelized wavelet revolutions for the recognition of R-wave. The algorithm has a high degree of immunity to noise and predictiveness.

Any kind of distortion in the reconstructed signal is introduced by the classical wavelet transformation that does not maintain the shift-invariant property. In 2010 [4], Pramodkumar contrasted the efficiency of DWT and SWT using three separate threshold methods for denoizing the ECG, namely universal, minimax threshold methods.

The proficient denoising calculation dependent on WAP utilizing round convolution in the wavelet area was shown by Marius Oltean [7] in 2006. The two changes are utilized for the exact recognizable proof of the ECG signal, to be specific Diversity Enhancement DWT (DE DWT) and restriction Invariant DWT (TI DWT). Zhao [8] recommended Empirical Mode Decomposition (EMD) with step channel to viably stifle power line recurrence noise without significant ECG signal bending, and B, Natwong, utilizing the wavelet entropy guideline to order individuals with and without ventricular late possibilities (VLP). The entropy is identified with the level of sign inconsistency. The higher the entropy, the more prominent the anomaly of the sign being referred to.

Hilbert–Huang Transform, comprising of Hilbert change calculation and scientific mode decay of arrangement of limited band signals got from the disintegration of the first sign dependent on prompt recurrence, was proposed by Yang [9] in 2008.

In the year 2009 [10], for the recognizable proof of boundary changes of the ECG signal because of unexpected body torment by reenactment measure, Fazlul Haque delineated and looked at the procedure utilizing FFT and WT. They indicated that WT delivers better results for nonfixed signals, for example, ECG.

In a request to analyze the measurable presence of the sign in the time-recurrence space, Sharma [11] recommended an ECG preparing approach zeroed in on higher request measurements for various wavelet subgroups. To decide the commotion amount in the sign, the kurtosis, and energy commitment of the subgroups are incorporated.

The proposed plot depends on a variable stage size firefly calculation (VSSFA) in the doubletree complex wavelet conspire, in which the VSSFA is utilized for improvement of the edge. This technique is tried by falsely applying white Gaussian sounds with blends somewhere in the range of 5 and 10 dB to a few customary and unpredictable ECG signals from the MIT/BIH arrhythmia information base.

Creator Mohammed Al Disi [13] recommended ECG signal reconstruction on the Internet of Things (IoT) Gateway and Compressive Sensing Efficacy Under Real-Time Constraints with a group. As an apparatus to drag out the battery life of clinical wearable gadgets, compressive detecting (CS) has been investigated. Anyway, at the deciphering end, it is normally connected with registering multifaceted nature, expanding the framework's inertness. Versatile processors are improving and all the more remarkable regarding registering. A neighborhood preparing approach that can conquer the deficiencies of distant sign handling is given by heterogeneous multi-course stages (HMPs). This paper represents the ongoing productivity of compacted ECG reproduction on the big. LITTLE HMP of ARM and the advantages they offer as the essential IoT engineering preparing unit. It likewise inspects the value of CS in limiting a wearable gadget's capacity utilization under continuous and equipment requirements.

A Georgani and others present various calculations in 2017 [15], including the Moving Average and Pan Tompkins calculations just as the utilization of wavelets to get highlights and attributes. Testing is completed at the last level by reproducing our program progressively records utilizing the TCP network convention to speak with a virtual sign source on the mobile phone.

7.2.2 Signal Processing Based on Artificial Intelligence

For the diagnosis of cardiovascular pathology, ECG analysis is an essential resource. Heart rhythm disturbances are evidence of underlying cardiovascular conditions in the ECG waveforms. It is challenging and time-intensive to manually interpret ECG signals and involves experts who are well trained to identify and categorize various waveform morphologies in the signal.

For heartbeat classification, multiple machine learning (ML) algorithms can be used, such as support vector machine (SVM), Bayesian classification, neural networks, LDA, k-nearest neighbor (kNN), and decision trees [16]. The classification model of the SVM is used to classify ECG beats in this section. SVMs find a boundary for judgment that maximizes the margin between the multiple groups. SVMs have outstanding generalization properties because they give sparse solutions expressed in a minimal number of support vectors (i.e., training samples that are closest to the decision boundary).

A significant principle in data science, which is used to enhance generalization and reduce overfitting, is the development of training and research datasets. The division of our data into two subsets is one way to produce test data: training data and testing data. The model is then fitted and checked on the unseen evaluation results using the training data. There are several ways to break the data that one would use. Two data splitting paradigms are widely used in the form of the ECG pulse classification: intra- and interpatient paradigms. The interpatient paradigm produces teaching and research subsets depending on the type of beat from the dataset. One of this system's key drawbacks is that it results in an over-optimistic estimation of the real success of the classifier. In both training and evaluation datasets, this is because heartbeats from the same patient will occur. The interpatient model uses a more robust split; here the patients used to train the classifier to vary from the patients used to test it [16].

Signal processing has brought us a bag of techniques that in the last 50 years have been refined and put to very good use. Autocorrelation, convolution, Fourier and wavelet transformations, Least Mean Squares (LMS) or Recursive Least Squares (RLS) adaptive filtering, linear estimators, compressed sensing, and gradient descent, to name a few are accessible. In order to solve various issues, various methods are used and often we use a mixture of these tools to create a device to process signals.

ML, or deep neural networks, is much easier to get used to and, regardless of the network model we use the basic mathematics are pretty straightforward. In the volume of data, they process to get the interesting findings, we currently have the sophistication and mystery of neural networks lie in. Figure 7.1 shows block diagram of a multilayer Convolution Neural Network (CNN) network [17, 19] in which all the stapes of processing like activation, pooling, vectorization, and fully connected layer shows the entire processing.

ML can do what signal processing can do but has greater sophistication naturally, with the advantage of being generalizable to multiple issues. In terms of complexity, the signal processing algorithms are ideal for the task but are unique to the specific issues they address. Instead of LMS or vice versa, we cannot use FFT, while we can use the same neural network processor, and only load a different set of weights to solve another task. The versatility in neural networks is that.

There are several issues in signal processing (philtre design) and ML (SVMs) from a computational viewpoint that can be formulated as problems of convex optimization. More broadly, optimization plays a significant role in

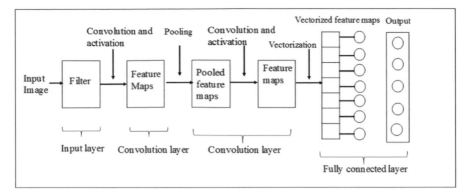

Figure 7.1   Block diagram of multilayer CNN networks.

both fields—in certain situations, the problem of optimization is to reduce any noise/uncertainty function subject to constraints that can be tuned, depending on the application. To a certain degree, in a logistic regression problem, you can loosely think of the problem of evaluating weights that optimize the likelihood function as similar to an analogous regression problem From a realistic viewpoint, to maximize prediction precision, it is always important to incorporate techniques from both signal processing and ML. For example, on wearable devices, signals from sensors containing data that correlates to human movement—the data may be very noisy, And classical signal processing techniques such as sample rate conversion and filtering need to be done to preprocess the data, which is then used for extracting features. As a consequence, the training data obtained depends on the filtering parameters used for pre-processing data, as shown in Figure 7.2 and it is necessary to consider these parameters when determining the feasibility of a predictive model that uses such training data.

Speech samples/recordings in the learning process should not be used as such. Sampling, cleaning (removal of noise or invalid samples, etc. or re-formatting the samples to an appropriate format may be needed for further processing. This phase is called "preprocessing of information."

We will also have to convert the data unique to the ML algorithm and the problem's understanding. Feature extraction on voice samples is done using signal processing in order to train the ML model to identify the patterns in the voice samples. In this case, pitch and Mel-Frequency Cestrum Coefficients (MFCC) [3] derived from the voice samples are the characteristics that are used to train the ML model.

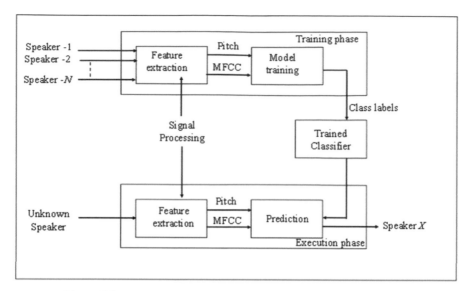

Figure 7.2   Speaker recognition using ML and signal processing [16].

The accessible dataset (set of input voice samples) is usually divided into two sets: one set for model testing and the other set for test uses (typically in 75–25% ratio). The training set is used to train the [16, 18] ML model and the test set is used to assess the ML algorithm's efficacy and performance. The training method should aim to generalize the fundamental relationship between the vectors of the function (input to the supervised learning algorithm) and the labels of the class (output of the supervised learner). One of the verification methods for determining the generalization potential of the ML model is cross-validation.

The preparation method should also stop overfitting, which in the implementation phase can cause poor generalization and erroneous classification [18]. We need to go back and make improvements to the previous steps if the output of the algorithm needs enhancement. In order to measure the efficacy, measures such as accuracy, recall, uncertainty matrix are usually used.

Recently, there is a growing interest in the creation of smart technologies and applications capable of communicating, such as the IoT and Human–Machine Interfaces, with their world. The word "smart" is castoff to designate a collection of unconventional features introduced using cultured algorithms of computational intelligence (CI). ML is an important field of CI that discovers the capability of computers/machines to acquireoverillustration, encoding, and storage of information. Related to the processing of the human brain, ML provides solutions to complex engineering issues.

The author [17] uses renowned image processing tools (Wiener filtering, 2D-DWT, and Probabilistic PCA) and ML models (Random Subspace Ensemble [RSE], K-nearest neighbors) to diagnose pathological brain images. In terms of accuracy, sensitivity, and precision for four datasets, the proposed approach was compared with 21 state-of-the-art algorithms. Another paper introduces a new semi-supervised active CNN algorithm capable of identifying SAR images in the training process without the need for a significant number of named samples. The approach initially applies active learning to the classified data, and then a semi-monitored phase of regularisation is programmed for the remaining unlabeled data.

ML is a computer science discipline that constructs mathematical models to understand data outlines to help people in their health-related decisions [18]. *Computer vision is a sub-field of ML which trains reproductions* of computers to quotation and analyze image data [19]. Several machine vision algorithms applied to the problems of object detection obtained better results than the best human success for classification accuracy [20]. The use of ML and computer vision algorithms (with the exception of the brain tumor segmentation challenge [21]) to interpret brain scan images is still unexplored.

A study on falling the number of input limitations of a multilayer perceptron neural network is one paper [22] on the implementation of intuitionist fuzzy inter requirements analysis. This would allow the weight matrices to be reduced as well as the neural network to be applied in small hardware, thereby saving training time and energy.

7.2.3 Problem Context

In recent years, the volume of imaging data has risen exponentially in medical imaging studies. This has made it more difficult for doctors to process the photos. They ought to read pictures with greater efficiency while retaining the same or better precision. Many Computer Assisted Diagnostic System (CADS) have been developed to diagnose various diseases using AI methods. There was a need to meet the following goals:

- To build an effective model for disease diagnosis.
- High-quality imaging helps doctors to make decisions even faster, and during the scanning process, thus speeding up and improving the precision of treatment.

7.3 Results and Discussion Based on Recent Work

We incorporated numerous approaches and counsel findings supported basic ways within the space of innovation of signal process since several studies administrated exploitation totally different algorithms here. One among the papers [23] discusses the utilization of reportable deep brain native field potentials (LFPs) for rigorous movement cryptography of patients with a brain disorder (PD) and Dystonia. For a precise estimate of finger motion and its coming laterality, a unique ensemble classifier was advised. The ensemble classifier consists of three classifiers of the bottom neural network, specifically feed-forward, radial-base, and probabilistic neural networks, whereas the majority-voting law was accustomed to fuse the ultimate decision-making selections of the three base classifiers.

The diagram indicates soft and hard limits above the denoized signal. In the MATLAB figure pane, the SNR and root mean square error (RMSE) are shown. As seen in Figure 7.3, the blue line is a signal free of noise and the denoted signal is the red line.

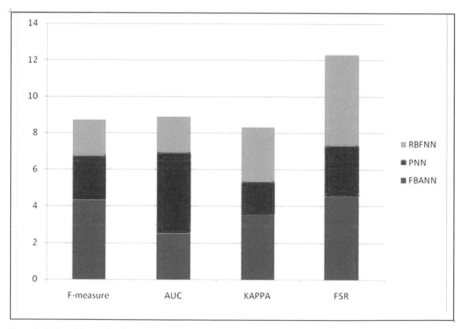

Graph 7.1　Numericalpresentation trials of base and ensemble classifiers to categorize inactive and left or right finger movement [23,26,28].

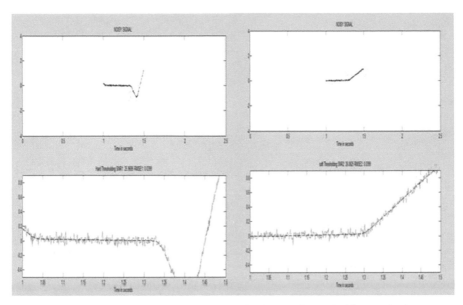

Figure 7.3 Denoized ECG signal by hard and soft thresholding method.

In the MATLAB figure pane, the SNR and RMSE are shown. As seen in Figure 7.4, the blue line is a signal free of noise and the denoted signal is the red line. The figure shows the denoted signal with the firm threshold.

In the MATLAB figure pane, the SNR and RMSE are shown. As seen in Figure 7.5, the blue line is a signal free of noise and the denoted signal is the red line.

As shown in Graph 7.2 and Table 7.1, the denoising effects of the noisy ECG signal were contrasted here using separate functions. The findings indicate that at noisy ECG signal denoizing, the shrinkage mechanism indicated in this job is very strong. It can not only achieve the highest SNR, but also preserve the denoized signal's similarity and smoothness. In truth, for all types of denoized signals, this feature can also be used. For the proposed approach, the increase of the SNR ratio shows that this is a powerful strategy for denoising nonstationary signals such as ECG signals.

When processing the ECG signal, the proposed threshold and shrinkage feature is helpful and improves the SNR ratio to obtain clean recordings and maintain the original signal form, particularly the peaks, without distorting the waves and segments. The key challenge is to recover from noisy recording and accomplish a true ECG signal by the proposed process.

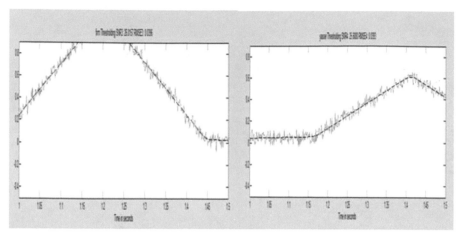

Figure 7.4   Denoized ECG signal by firm and Yasser thresholding method, blue is a noisy free signal, and red is denoized signal.

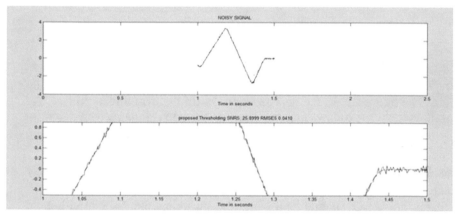

Figure 7.5   Denoized ECG signal by hybrid thresholding method, blue is a noisy free signal and red is denoized signal.

7.4 Real-Time Applications

1. **Speech Analysis**—A profoundly ingrained and instinctive mode of communicating for humans is to use voices to access information and to communicate with the world. Speech recognition—the translation of speech audio into word sequencesis a requirement for any conversation dependent on speech. The conditional independence presumption of the performance goals suffered by the conventional HMM-based phone state

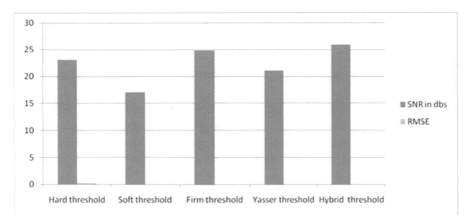

Graph 7.2   Different thresholding methods with SNR and RMSE.

Table 7.1   Both SNR and RMSE are better than the other four methods due to good similarities and smoothness [5, 6, 7, 9, 10]

Method	SNR in dbs	RMSE
Hard threshold	23.06	0.268
Soft threshold	17.07	0.020
Firm threshold	24.84	0.029
Yasser threshold	21.05	0.0285
Hybrid threshold	25.90	0.0487

modeling is no longer relevant with the implementation of RNNs for speech modeling, and the research area moves to the more flexible full sequence-to-sequence models. A strictly neural sequence-to-sequence model, such as CTC and LAS [24], has been of great interest in learning.

2. **Music Analysis**—Music records generally include a greater spectrum of sound sources of interest relative to voice. In certain forms of music, their occurrence in terms of time and frequency meets traditional restrictions, establishing dynamic dependencies within and between sources. This opens up a wide variety of possibilities for explaining music recordings automatically [21, 22]. Deep learning has been successfully extended to various music processing tasks and drives industrial applications with automated explanations for large catalog searching, content-based music suggestions in the absence of user details, and even profane chords for a song to play together with automatically derived chords. There is, however, an agreement about what input representation

to use (log-Mel spectrogram, constant-Q, raw audio) and what architecture to use (CNNs or RNNs or both, 2D or 1D convolutions, small square or large rectangular philters) on the study side, either inside or through activities, leaving various unanswered questions for upcoming research [24, 25].

3. **Separation of Environmental Sounds**—Certain sounds also hold a wealth of relevant knowledge about our surroundings, in addition to voice and music cues. There are many applications for computer processing of ambient sounds, such as context-aware systems, acoustic surveillance, or multimedia indexing and retrieval. It is usually achieved with three simple approaches: (a) recognition of acoustic scenes, (b) identification of acoustic events, and (c) marking. Using a multilabel classifier to jointly forecast the behavior of several classes at once has been shown to yield improved results instead of using a single-class classifier independently for each class. For example, this may be because the multiclass classifier is able to model the interaction between groups that are simultaneously involved [22, 25].

4. **Localization and Tracking**—Multichannel audio helps sound sources to be located and monitored, that is to determine their spatial positions and monitor them over time, and can be used for example, as part of source isolation or speech amplification method to distinguish a source from the approximate source direction or to estimate the behavior of multiple speakers in an idolization system. In general, source localization involves the use of interchannel input, which can also be learned from within-channel features by a deep neural network with a suitable topology, for example by convolution layers where several channels are spanned by the kernels [21, 23, 24].

5. **Audio Enhancement**—Speech enhancement strategies are aimed at enhancing voice production through noise reduction. Conventional denoizing solutions, such as Wiener techniques, generally consider static noise, where time-varying noise can be modeled as deep learning approaches. Different forms of networks were analyzed for progress in the literature, such as denoising Autoencoders [22, 23], recurring networks, and coevolutionary networks.

6. **Modeling**—Rather than interpreting the signs, we can model them. That is, we should think of the coarse-grained signal as a series of latent variables rather than decomposing a fine-grained signal into a coarse-grained signal, and then use the fine-grained signal to statistically infer them. As it deals with data from multiple resolutions, this is considered a multi-revolutionary model [23, 25].

7.5 Conclusion

Medical info, which can benefit millions of lives around the globe, is one of the various areas where audio analysis can be implemented. By providing a data goldmine to the analysts, this phase will carry healthcare to another level.

The very first effort of its kind was to decide whether there were any signs of cancer in it by an IBM group who used prior reported blood samples. The ECG signal signals the heart's electrical function. Variations in ECG signal amplitude and length from a predefined sequence have been observed.

Routinely used for heart abnormality diagnosis. Owing to the difficulties of personally interpreting these differences, a computer-aided diagnostic device may assist in tracking the state of cardiac health. Nonlinear extraction approaches are good candidates for extracting information from the ECG signal because of the nonlinear and nonstationary existence of the ECG signal. Because artificial neural networks are fundamentally a pattern matching technique based on nonlinear input-output mapping, morphological changes in nonlinear signals such as the ECG signal can be effectively detected.

It seems like too many proposed devices are indeed very flexible. While it emerged as a signal separating technique, its application has grown well beyond this field; it was used relatively early for concerns such as noise reduction in the detection of ECG signals, clinical ECG analysis. By using an enhanced algorithm, the speed of execution of the algorithm can be increased. The ECG signal study has interfered with the aid of the suggested approach that will essentially denoize the signal distorted by different noises, different noises may be denoized.

In pattern classification, the problem of choosing an appropriate collection of applicable features plays an important role. If we have the right pattern classification scheme, it is not sufficient to gain greater consistency in pattern classification. The chosen features must be capable of at least a useful degree of differentiation between the groups. They will have become obsolete. It is important to screen the selected features for redundancy and irrelevance. While various methods can be used to extract various features from the same raw data, it is fundamentally important to combine a feature extractor and a pattern classifier.

Different suggested methods for function extraction, diagnosis of cardiac wellbeing, and prediction of ischemia from ECG signals have been addressed here the procedure has advanced a lot today. To learn how radiation can be more strengthened, Google Deep Mind is dealing with cancerous tissues. In healthcare, this form of cutting-edge application is precisely the right place to begin using signal processing in ML. ML algorithms that have historically been developed to accommodate signal processing operations such as convolution and FFTs are implemented on hardware and embedded applications (such as DSP chips). There are programs (such as exercise wearables) in which when the hardware performs BOTH signal processing and ML processes, latency and power consumption will need to be considered. In summary form, we can conclude that signal processing can be used to prepare data for algorithms for ML (such as signal cleaning and speech recognition extraction features). ML will be used to address certain difficult problems of signal processing (such as using ML to predict speech activity). There would also be a greater interdependence between signal processing, ML, and the design of low-level software in resource-constrained computer applications.

Acknowledgment

The authors would like to thank the Sreenidhi institute of science and technology, Hyderabad for providing the infrastructure to conduct the research.

References

[1] S.A. Choukari, F. Bereksi-Reguing, S. Ahmaidi and O. Fokapu, "ECG signal smoothing based on combining wavelet denoising levels". Asian Journal of Information Technology 5(6), pp 666–77, 2006.

[2] C. Saritha, V. Sukanya, Y. Narasimha Murthy, "ECG Signal Analysis Using Wavelet Transform", PACS number: 87.85.J; 02.30.Nw February 2008.

[3] Abdel-Rahman Al-Gawasmi and Khaled Daqrouq," ECG Signal Enhancement Using Wavelet Transform" ISSN: 1109-9518 Issue 2, Volume 7, April 2010

[4] Pramodkumar, Dewanjali Agnihotri, "Biosignal Denoising via Wavelet Thresholds", IETE journal of Research, vol 56, issue 3, pp. 132—138, May–June 2010

[5] Seung Min Lee, Ko Keun Kim, Kwang Suk Park Wavelet Approach to Artifact Noise Removal from Capacitive Coupled Electrocardiograph 978-1-4244-1815-2/08/$25.00 © 2008 IEEE. 2944.

[6] F. Abdelliche, A. Charef, M.L. Talbi, M. Benmalek, "A Fractional wavelet for QRS Detection", 0-7803-9521-2/06, IEEE pp.1186-1189, 2006

[7] Marius costache, dumitrutoader, mariusoltean, AlexandruIsar," MAP filtering in the Diversity-enhanced wavelet domain Applied to ECG signals Denoising", ICASSP pp. II-1196-1199, 2006

[8] Zhi-Dong Zhao, Yang Wang, Analysis of Diastolic Murmurs for Coronary Artery Diseasebased on Hilbert Huang Transform, Machine Learning and Cybernetics, Volume: 6, 2007.

[9] Xiao-Li Yang, Jing-Tian Tang, Hilbert-Huang Transform and Wavelet Transform for ECG Detection, Wireless Communications, Networking and Mobile Computing, 2008.

[10] A.K.M. FazlulHaque, Md. HanifAli, M. Adnan Kiber and Md. Tanvir Hasan, "Detection of Small Variation of ECG Features using Wavelet", ARPN Journal of Engineering and applied Sciences, vol. 4, No. 6, pp. 27–30, August 2009

[11] L.N. Sharma, S. Dandapat, A. Mahanta, "ECG Signal Denoising using *Higher Order Statistics in Wavelet Sub bands*", *Biomedical signal Processing and Control*, Elsevier, vol. 5, pp. 214–222, 2010

[12] VinuSundararaj, An Efficient Threshold Prediction Scheme for Wavelet Based ECG Signal Noise Reduction Using Variable Step Size Firefly Algorithm, International Journal of Intelligent Engineering and Systems, Vol. 9, No. 3, 2016

[13] Mohammed Al Disi, Hamza Djelouat, Christos Kotronis, Elena Politis, Abbes Amira And Guillaume Alinier4, ECG Signal Reconstruction on the iot-Gateway and Efficacy of Compressive Sensing Under Real-Time Constraints, IEEE ACCES, VOLUME 6, 2018.

[14] VinuSundararaj, Optimised denoising scheme via opposition-based self-adaptive learning PSO algorithm for wavelet-based ECG signal noise reduction,

[15] International Journal of Biomedical Engineering and Technology, Volume 31, Issue 4 ,2019

[16] A Georganis, N Doulgeraki, and P Asvestas, a real time ECG signal processing application for arrhythmia detection on portable devices, Journal of Physics: Conference Series, J. Phys.: Conf. Ser. 931 012004, 2017.

[17] By Mathura Nathan, Introduction to Signal Processing for Machine Learning, gaussian waves, Word Press, 2020

[18] Martin Cenek, Masa Hu, Gerald York and Spencer Dahl , Survey of Image Processing Techniques for Brain Pathology Diagnosis: Challenges and Opportunities, Front. Robot. AI, 02 November 2018.

[19] Alpaydin, E. Introduction to Machine Learning. Cambridge, MA: MIT Press,2014

[20] Shirai, Y. Three-Dimensional Computer Vision. Berlin; Heidelberg: Springer Science & Business Media,2012

[21] Roszkowski, O., Deng, J., Su, H., Krause, J., Satheesh, S., Ma, S., et al. Imagenet large scale visual recognition challenge. Int. J. Comput. Vis. 115, 211–252. Doi: 10.1007/s11263-015-0816-y,2014

[22] Menze, B., Jakab, A., Bauer, S., Kalpathy-Cramer, J., Farahani, K., Kirby, J., et al. "Multimodal brain tumour image segmentation. Benchmark: change detection." In Proceedings of MICCAI-BRATS 2016 (Athens: MICCAI), 2016

[23] Sotir Sotirov, Vassia Atanassova, Evdokia Sotirova, Veselina Bureva1 Deyan Mavrov1, Application of the Intuitionistic Fuzzy intercriteria Analysis Method to a Neural Network Pre-processingProcedure, the authors – Published by Atlantis Press, 2015.

[24] Mohammad S. Islam, Khondaker A. Mamun, and Hai Deng, Decoding of Human Movements Based on Deep Brain Local Field Potentials Using Ensemble Neural Networks, Computational Intelligence and Neuroscience Volume 2017.

[25] Hendrik Purwins, Bo Li, Tuomas Virtanen, Jan Schlüter, Shuo-yiin Chang, Tara Sainath, Deep Learning for Audio Signal Processing, Journal of Selected Topics of Signal Processing, Vol. 14, No. 8, May 2019.

[26] Mesaros, T. Heittola, E. Benetos, P. Foster, M. Lagrange, T. Virtanen, and M. D. Plumbley, "Detection and classification of acoustic scenes and events: Outcome of the DCASE 2016 challenge," IEEE/ACM Transactions on Audio, Speech, and Language Processing, vol. 26, no. 2, pp. 379–393, 2018.

8

Reversible Image Data Hiding Based on Prediction-Error of Prediction Error Histogram (PPEH)

K. Upendra Raju[1], D. Srinivasulu Reddy[1], and P.E.S.N. Krishna Prasad[2]

[1]ECE, Sri Venkateswara College of Engineering, Tirupati-517507, India
[2]CSE, GIT, GITAM Deemed University, Visakhapatnam-530045, India
E-mail: kupendraraju@gmail.com; cnudega@gmail.com;
surya125@gmail.com

Abstract

Reversible data hiding (RDH) or lossless data hiding techniques has been initially studied in the family of signal processing. It also referred as invertible or lossless data hiding is to embed the information into a cover signal to make a marked data (Stego data). From the marked data, the original signal can be exactly recovered after extracting the embedded data. In this chapter, we propose a unique reversible data hiding technique based on prediction-error of prediction error (PE) histogram (PPEH) of an element to reversibly carry the key data. In the projected methodology, the pixels to be embedded are firstly predicted with their neighboring pixels to get the PEs because most natural pictures perpetually contain edges, it's not appropriate to predict these pixels using existing prediction ways. For more precise prediction, two prediction ways are adaptively used to calculate PE consistent with the characteristic of a pixel. And, a sorting technique based on the local complexity of a pixel is employed to gather the PPEs to get an ordered PPE sequence so that, smaller PPEs are processed initial for data embedding. By reversibly shifting the PPEH with optimized parameters, the pixels equivalent to the altered PPEH bins can be finally changed to carry the complete secret data. Experimental results have implicit that, the proposed algorithmic program can benefit from the prediction procedure, sorting technique as well as parameters choice.

Keywords: Reversible data hiding, prediction-error of prediction error (PPE), sorting, adaptive, watermarking.

8.1 Introduction

With the increased usage of internet, data communication through media has become easier and more popular. Digital media is the major source of communication for transmitting the data. When secret data of persons, organizations, and corporate are communicated through network in every moment, which entails a risk of copy or corruption when data are received. In this computer world, it is very important to provide security for the personal and private information, and protect the copyrights of data. Hence, there is a need to provide security for safe transmission of data on the chosen communication channel [1].

Security of the data can be classified into three broad categories, such as cryptography, watermarking, and steganography. The data can be encrypted by using encryption key to get the cipher text and the same key used to decrypt the cipher text into plain text. But security level is low in this technique, as the attacker can destroy and try to reconstruct the original text by using hackers own algorithms.

The secret data are hidden in original data by watermarking techniques. The watermark is imperceptible and creates invulnerability against various types of attacks.

Steganography is an advanced technique in art and science of invisible communication. The word Stego means covered or hides and graphy means writing or drawing. The information is completely secure and completely undetectable, and also prevents generating suspicion to the transmission of a hidden data [2].

Embedding is the process of hiding the information, also referred to as text, copyright, or information in the transmitting media for the use of copy control, content authentication, distribution pursuit, and broadcast monitoring [3, 4]. In most instances of information stowing away, due to the presence of the concealing information, the host media can have unavoidable ability to some contortion that cannot be redressed. Despite the fact that the twisting resulting from information stowing away can be made imperceptible to the human visual framework, it might at present be inadmissible for a couple of sorts of pictures in extraordinary applications, similar to therapeutic pictures, military pictures, remote detecting pictures, and plan safeguarding. In such conditions, reversible information concealing strategies are intended to infix information into a photo in a reversible way, which is furthermore expressed as lossless or invertible information implanting that implies that the host media can be precisely recovered.

For past two decades, data hiding has emerged as an interesting area of research [5]. In this method, secret data can be encoded into a cover medium, and subsequently facilitates the user to take out the embedded data from the Stego medium for various applications. In many data hiding methods, the data are distorted during the operation and prevent the receiver from retrieving original form of data. In view of the above difficulties, reversible data hiding (RDH) also called lossless or invertible data hiding is proposed. Recently, there are many reversible data hiding methods that have been pronounced in the literature survey, which are used for secret communication [6].

Early reversible information concealing calculations basically fixated on lossless pressure [7–10]. Fridrich et al. proposed many algorithms using compression of bit-planes [7] and vector states [8] for better performance. Celik et al. proposed a lossless generalized-LSB data hiding technique [9, 10] that proposed to compresses a group of selected features from a picture and embeds the payload within the space saved by the compression. Lossless recovery of the first media is accomplished by compressing bits of the signal that are responsible to embedding distortion and transmitting these compressed descriptions as a part of embedded payload. Usually, these methods have low embedding capacity with high distortion.

Several approaches to reversible watermarking [11–14] are often classified the techniques based on histogram. The message is embedded into the histogram bin using peak/zero bins within the histogram image were proposed by Ni et al. [11]. In [13], the histogram technique was unlimited that using pixel variations to extend the hiding capability. A disadvantage of [13] is that severe artefacts in embedded image caused by the rough image pixel value prediction technique. Recently, a new histogram modification primarily based reversible data hiding algorithmic rule considering the human visual system is existing in [14]. It uses the idea of difference threshold [15, 16] to cut back the perceptual distortion. Based on integer Haar wavelet transform, Tian [17] proposed another scheme using the Difference-Expansion (DE) embedding. This scheme calculates the variations of neighboring pixel values, and selects appropriate distinction values to fix data. This scheme typically achieves higher embedding capability and lower distortion, such a large amount of its variants and extensions have been proposed within the literature. Fallahpour et al. [18] improved the location map compression by sorting possible expandable locations and wanted to mechanically produce just enough capability to embed the required payload, therefore keeping the distortion minimal. Li et al. [19] formed better use of the redundancy of neighboring pixels. Hu et al. [20] targeted on raising the overflow location map, therefore achieving sensible compressibility.

The existing RDH algorithms, in most cases, predict the pixels with a well-designed predictor initially. Then, the key bits are embedded in the resultant prediction-error histogram (PEH). The pixel prediction procedure is usually needed to well predict the pixels due to the abstraction correlations between neighboring pixels; the present methods usually exploit the DE [17] of a pixel to hold the key data. In fact, there should also exist robust correlations between neighboring PEs. One proof can be found in the prediction mechanism of recent video lossy compression. For instance, in intra prediction, the prediction block for an intra 4×4 luma macroblock (consisting of 16 pixels) will be generated with 9 attainable prediction modes due to the spatial correlations between neighboring pixels. Then, so as to enhance the coding potency, the prediction mode of a luma macroblock should be further predicted from the prediction modes of neighboring luma macroblocks since correlations additionally exist between the neighboring prediction modes based on this perspective.

We intend to propose a reliable PPEH-based RDH techniques in this chapter, in which the PEs of the pixels to be embedded are firstly determined consistent with their neighboring pixels to reduce the distortion for a required payload; the pixels are sorted by the native complexities. A pixel with a lower complexness is preferentially embedded with a secret bit. By shifting the resultant PPEH, the corresponding pixels can be finally altered to carry the key data. Experiments have implicit that the projected technique will offer good payload-distortion behavior.

Different researchers are worked on reversible data hiding and steered several ways using different technologies and additionally research goes on to enhance more and more.

8.2 Existing Methodology

The proposed technique uses grayscale images. The information embedding procedure chiefly consists of four parts: the pixel prediction, the prediction of PE, the utilization of pixel sorting, and the data hiding in the image. In addition to the prediction of PEs, a pixel sorting technique, also known as pixel choice, is used. The purpose is to place the PPEs in a decreasing order of the prediction accuracy so that smaller PPEs are often processed initial, which edges the embedding performance.

8.2.1 Histogram-Based RDH

The first proposed [8] method is the RDH technique based on histogram. In this method, the messages are embedded based on the minimum points and maximum points utilized by Ni et al. [11] 2006, for the given histogram image. The highest bins of the histogram give the amount of embedding capacity that gives the number of pixels. Figure 8.1 shows the corresponding embedding process. This method is restricted to two main factors. The first factor is that in the first place to have empty bins in the required histogram. Otherwise, we follow the center plot as shown in Figure 8.1 But, this algorithm fails if Figure 8.1 cannot happen. Although the thinnest possible bins of histogram by generating empty bins to increase the likelihood. There is no guarantee that vacant bin will always appear. Besides, this method gives

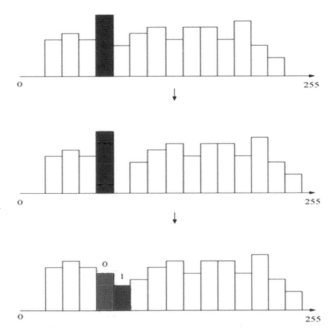

Figure 8.1 The highest bin in an image histogram is first found (the red bin), then all the bins on one side of it (right side in this case) is moved by one bin to leave an empty bin beside the highest bin (middle plot). Finally, the pixels in the red bin are coded by the embedding message so that they form two new adjacent bins (the yellow and orange bins).

the less embedding capacity because it produces a smaller number of pixels in the highest bin. In case, if an empty bin is not available, then all the pixel information is kept in the lowest bins. But the embedding capacity is lower very largely. The embedding capacity is lower is the second limiting factor of this method. We observe that the maximum number of bits that can be embedded into an image is the number of pixels in the highest bin in the histogram. Although this approach generally produces very little image distortion, which is due to the fact that most pixel values do not change and only a small portion of the pixel values gets changed by just 1, the limited embedding capacity still hinders its practical employment.

8.2.2 PEH-Based RDH

Hong et al. [9] was first proposed this approach that the histogram of the PE twice is applied the method in [8]. Figure 8.2 shows the data embedding process based on highest bins. But still this method has some drawbacks,

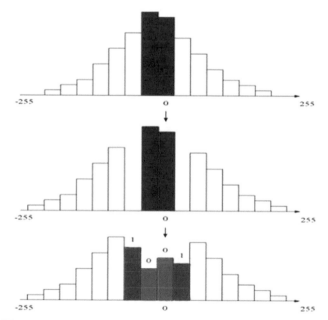

Figure 8.2   The two highest bins, usually the middle two, in the prediction error histogram are first selected (red bins in the top plot). Then the left and right parts beside these two bins are moved to the left and right by one bin, leaving two empty bins just beside the two red bins (middle plot). Finally, the pixels in the left red bin are coded/spilled into its left empty bin and independently the same is done for the pixels in the right bin.

and these results have two major advantages. The initial advantage is that the PE reaches the maximum possible value of 255, it is almost always possible to move the left and right portions of the histogram. The second advantage is that, instead of using one bin for data embedding as in Figure 8.1, two highest bins are used for data embedding. It has been observed that with most commonly used prediction methods, the PE is more concentrated near 0. Therefore, generally, the heights of the red bins in Figure 8.2 are bigger than the height of the red bin in Figure 8.1, which means the embedding capacity of the new approach more than doubles.

8.3 Proposed Method

We proposed a general data embedding methods in the PE domain through histogram modification. This can be proposed work from the past work using histogram bins. In the following, we explain the proposed algorithms development and prove that some data embedding ways, as well as the DE approach, which is one of the most effective in performance, are special cases during this framework.

For pixel prediction, all pixels are divided into two sets: the cross set and dot set Figure 8.3. The cross set is used for data hiding and dot set for pixel prediction. We use this rhombus pattern as it can maintain a good prediction performance [10]. It is noted that, one may use other efficient prediction patterns. We consider $u_{i,j}$ in Figure 8.3 as an example. In Figure 8.3, $u_{i,j}$ will be predicted from its four neighbors using the interpolation operation [1].

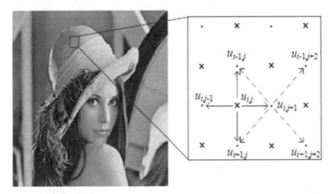

Figure 8.3 The pixel prediction pattern. The pixel $u_{i,j}$ will be predicted from its four neighbors in the dot set, and $u_{i,\,j+1}$ will be predicted from its four neighbors still in the dot set.

The resultant PE is computed as

$$e_{i,j} = u_{i,j} - u_{j,j+},\qquad(8.1)$$

where $u_{j,j+}$ is the prediction value. The prediction of $e_{i,j}$, denoted by $e_{i,j|}$ will be the average of the PEs of neighboring pixels in the dot set. As shown in Figure 8.3, the neighbor $u_{i,j+1}$ will be predicted from four neighbors still in the dot set using the interpolation operation. Therefore, we can finally obtain the PE of $e_{i,j}$, namely, $e_{i,j+} = e_{i,j} - e_{i,j|}$. After sorting, we collect a part of the PPEs with relatively smaller values to generate a PPEH. We choose two peak-zero bin-pairs (l_p, l_z) and (r_p, r_z) for data hiding.

During data embedding, the PPEs in range $(l_p, l_z) \bigcup (r_p, r_z)$ are shifted to avoid ambiguous. For a PPE with a value of l_p or r_p, if the secret bit equals "0," the PPE will be kept unchanged; otherwise, it will be modified as (l_{p-1}) or (r_{p+1}). Thus, a marked PPEH can be generated and the used pixels can be finally modified with $\{\pm 1, 0\}$ operation to match the PPEH. This way, the marked image can be finally constructed. It is noted that, there is no need that $h(l_p) + h(r_p)$ is maximal as long as $h(l_p) + h(r_p)$ is larger than the size of the required payload. Here, **h(x)** denotes the frequency of the PPEH bin x. It is desirable that, the expected number of altered pixels is as small as possible in order to keep the distortion low as a larger number of modified pixels usually correspond to a higher distortion. In applications, $|l_z|$ and $|r_z|$ are both small, indicating that, the computational cost to find the two peaks-zero bin-pairs will be very low.

Furthermore, as altering a pixel may result in overflow/underflow problem, to ensure reversibility, the boundary pixels in the cross set should be shifted in advance and then recorded to produce a location map, which should be self-embedded with the secret data. Since the boundary pixels in nature images are relatively rare, the effect on the pure payload could be ignored.

Note that, since changes in the cross set will not affect the dot set, the dot set can be applied for data hiding after data hiding with the cross set. The advantage is that, when using only the cross set, pixels with larger PPEHs have to be modified to carry the required payload; while, for the consecutive usage of the cross set and dot set, two sets of sorted PPEs with smaller values can be used first, implying that, the required payload of each set is approximately half of that for data embedding only with the cross set, and thus could maintain a lower distortion. After the receiver acquires the marked image, he can completely extract the hidden data and recover the original image without loss. It can be performed with an inverse operation to the data hider side.

8.4 Results and Discussions

We have implemented the proposed PPEH-based RDH algorithm, and applied it to the standard testing images (512 × 512, and 8-bit greyscale) (a) Baboon and (b) Lena. The payload-distortion performance was evaluated by comparing with existing system [9] and the obtained results are shown below. The mean square value (MSE) between the original image and stego image is defined as

$$MSE = \frac{1}{M \times N} \sum_{i=0}^{M-1} \sum_{j=0}^{N-1} (x_{i,j} - x_{i,j}^i)^2 \qquad (8.2)$$

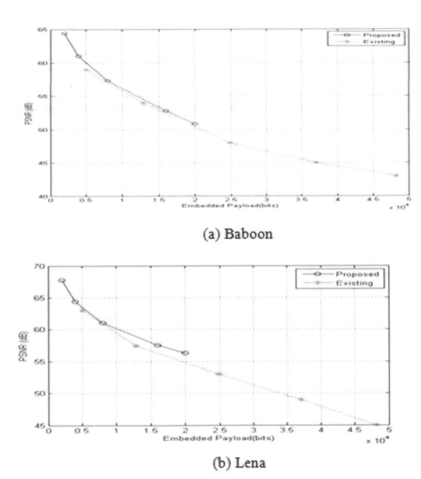

(a) Baboon

(b) Lena

Figure 8.4 The payload-distortion performance comparison between the state-of-the-art methods of existing and the proposed method.

The peak signal to noise ratio (PSNR) is to evaluate the quality of the stego image and is defined as

$$PSNR = 10 \times \log_{10} \frac{255^2}{MSE} \qquad (8.3)$$

It is observed from Figure 8.4 that, the proposed method outperforms these state-of-the-art works in terms of the payload distortion performance of different images like Baboon and Lena. The PSNR value is 48 dB in existing system and 65 dB in the proposed system for baboon image. For Lena image the PSNR value is 68 dB in proposed system which is higher than the existing system which is only 63 dB.

8.5 Conclusion

In this chapter, an efficient RDH scheme based on PPEH is proposed. The pair wise PPEH is a novel reversible mapping that utilizes the correlations among PEs. With the help of this type of correlations, the distortion can be controlled at a low level, and thereby the proposed scheme outperforms some state-of-the-art RDH algorithms. In future, there is still room for further improvement such as by designing a better predictor, better evaluating local complexities, or applying better data embedding operation. And, it is possible to apply the PPE for different cover media, and/or employ high-order PEs for data hiding.

References

[1] J. Fridrich. 2009. Steganography in Digital Media: Principles, Algorithms, and Applications. Cambridge, U.K.: Cambridge Univ. Press.

[2] T. Morkel, J.H.P. Eloff and M.S. Olivier. An Overview of Image Steganography.

[3] I. Cox, M. Miller, J. Bloom, J. Fridrich, and T. Kalker, Digital Watermarking and Steganography, Second Edition. Morgan Kaufmann Publishers Inc., San Francisco, CA, USA, 2007.

[4] D. M. Thodi and J. J. Rodriguez, "Expansion embedding techniques for reversible watermarking," IEEE Trans. on Image Processing, vol. 16, no. 3, pp. 721–730, 2007.

[5] J. Fridrich. 2009. Steganography in Digital Media: Principles, Algorithms, and Applications. Cambridge, U.K.: Cambridge Univ. Press.

[6] K. Upendra Raju, N. Amutha Prabha "A Review of Reversible Data Hiding technique based on Steganography", Proceedings of ARPN Journal of Engineering and Applied Science, vol.13, No.3, page no. 1105–1114, Feb. 2018, ISSN. 1819-6608.

[7] J. Fridrich, M. Goljan, and R. Du, "Invertible authentication," In Proc. of the SPIE Security Watermarking Multimedia Contents, pp. 197–208, San Jose, CA, USA, 2001.

[8] J. Fridrich, M. Goljan, and R. Du, "Lossless data embedding—New paradigm in digital watermarking," EURASIP Journal on Applied Signal Process, vol. 2002, no. 2, pp. 185–196, 2002.

[9] Wein Hong, Tung-Shou Chen, and Chin-Wei Shiu, "Reversible Data Hiding Based on Histogram Shifting of Prediction Errors"2008 international Workshop on Geoscience and Remote Sensing, IEEE Computer Society, pp. 578–581, 2008.

[10] M. U. Celik, G. Sharma, and A. M. Tekalp, "Lossless watermarking for image authentication: A new framework and an implementation," IEEE Trans. on Image Processing, vol. 15, no. 4, pp. 1042–1049, 2006.

[11] Z. Ni, Y. Q. Shi, N. Ansari, and W. Su, "Reversible data hiding," IEEE Trans. on Circuits and Systems for Video Technology, vol. 16, no. 3, pp. 354–362, 2006.

[12] M. Fallahpour and M. H. Sedaaghi, "High capacity lossless data hiding based on histogram modification," IEICE Electronics Express, vol. 4, no. 7, pp. 205–210, 2007.

[13] W.-L. Tai, C.-M. Yeh, and C.-C. Chang, "Reversible data hiding based on histogram modification of pixel differences," IEEE Trans. on Circuits and Systems for Video Technology, vol. 19, pp. 906–910, 2009.

[14] S.-W. Jung, Le Thanh Ha, and S.-J. Ko, "A new histogram modification based reversible data hiding algorithm considering the human visual system," IEEE Signal Processing Letters, vol. 18, no. 2, pp. 95–98, 2011.

[15] A. M. Alattar. 2004. Reversible watermark using the difference expansion of a generalized integer transform. IEEE Trans. Image Process. 13(8): 1147–1156.

[16] J. Tian, "Wavelet-based reversible watermarking for authentication", Proc. SPIE, vol. 4675, page no. 679–690, Apr. 2002.

[17] J. Tian. 2003. Reversible data embedding using a difference expansion. IEEE Trans. Circuits Syst. Video Technol. 13(8): 890–896.

[18] M. Fallahpour and M. H. Sedaaghi. 2007. High capacity lossless data hiding based on histogram modification. IEICE Electron. Exp. 4(7): 205–210.

[19] X. Li, B. Li, B. Yang and T. Zeng. 2013. General framework to histogram shifting- based reversible data hiding. IEEE Trans. Image Process. 22(6): 2181–2191.
[20] Yongjian Hu, Heung-Hyu Lee, and Jianwei Li, DE Based Reversible Data Hiding with Improved Overflow Location Map, IEEE Trans. Trans. Circuits Syst. Video Technol. 19(2): 250–260.

9

Object Detection using Deep Convolutional Neural Network

G.A.E. Satish Kumar and R. Sumalatha

Electronics and Communication Engineering, Vardhaman College
of Engineering, Hyderabad, India
E-mail: gaesathi@gmail.com; amrutha.suma18@gmail.com

Abstract

Object detection is a challenging task in computer vision applications that includes person detection, face detection, vehicle detection, pedestrian's detection, and human behavior detection. In conventional object detection methods, the features are extracted using handcrafted algorithms. The main aim of the object detection is to identify the object's location in the images or videos, and also gives the information about class of an objects. But with the different poses of person, occlusions and illumination changes of images, it is very difficult to detect the objects in the images. Due to recent developments in technologies, deep learning methods are used to detect and classify the objects in computer vision applications. In this chapter, we have been presented a survey on object detection algorithms. Various datasets like PASCAL VOC versions, ILSVRC, MS COCO, and Open of object detection (OICOD) provide information. In the experimental analysis, various methods are compared and finally discussed the advantages and disadvantages of the different object detection methods. Finally, many interesting directions are presented for future work in an object detection system.

Keywords: Object Detection HOG, LBP, SURF, SSD, YOLO, RCNN

9.1 Introduction

In computer vision applications, object detection, localizing, and recognizing in the images and videos are a huge problem. Feature extraction play an important role in computer vision applications. Many researchers have been suggested different algorithms for object detection including SIFT, SURF, HOG, and Haar [1–5]. These algorithms are extracted either low level features or high-level features and also these are not robust to illumination changes and pose variations. Nowadays, deep convolutional neural networks have become most powerful tool for object detection. Because it automatically extracts low level, middle-level and higher-level features from an image. Object detection provides a number of applications, including face detection, face recognition, vehicle detection and tracking, content-based image retrieval, and video surveillance [6, 7]. Finally, objects are classified using classifier algorithms like SVM [8], KNN [9], and Ada boost [10].

9.2 Related and Background Work

Zhu et al. suggested a deep convolutional neural network (DCNN) for aerial object detection by taking account into orientations and the segmented images are classified using support vector machine (SVM) [11]. Anuj Mohan et al. proposed example-based detectors algorithm for detecting and classifying the people, and also robust for detecting people in low-contrast images [12]. Zeng et al. have been presented an overview of place recognition and discussed about CNN, bag-of-visual words, and VLAD [13]. Josifovski et al. have been proposed a novel approach for detecting the objects and 3D pose estimation using CNN [14]. Mohana et al. have been implemented one algorithm for detecting and tracking the vehicles in traffic, and yolov3 is used to detect multiple objects of KITTI and COCO database images [15]. Girshick et al. proposed an algorithm for object detection on PASCAL VOC 2012 and achieves more than 50% improvement in mean average precision [16]. Cheng et al. implanted an algorithm for object detection using faster RCNN [17]. Zhou et al. suggested a Faster RCNN algorithm for object detection on VOC2007 dataset and achieves 73.2% map and this value is higher than the conventional methods like RCNN, SPP Net, and fast RCNN [18]. Masita et al. discussed various methods proposed by different researchers and also given the information about the recent advancements in object detection [19]. Zhang et al. proposed a multiobject detection by fusing the CNN features using multi-branch group fusion on RGB-T relevant object detection datasets, and it provides good improvement in poor illumination, composite background, and low contrast [20]. Suhail et al. presented a review on different object detection approaches using convolutional neural network [21]. Ahmad et al. proposed a modified YOLOV1 algorithm for object detection effectively and archives 58.5% average recognition rate on VOC 2012 and 65.6% on VOC 2007 dataset [22]. Gu et al. proposed an algorithm to detect the body parts and also proposed adaptive joint Non-Maximum Suppression algorithm to improve the precision and recall rate while detecting the pedestrians CUHK-SYSU Person Search Dataset [23]. Liu et al. discussed the recent developments in object detection using several systems with deep learning methods [24]. Szegedy et al. presented DNN base object mask recognitions algorithm using a multiscale course-to-fine procedure [25]. Zhao et al. discussed recent developments in object detection using deep learning techniques and also provides comparison between various algorithms of current technologies with conventional methods [26]. Wang et al. proposed an algorithm for detecting the high-resolution remote sensing images by fusing object and scene image features with deep convolutional neural network [27]. Haidar et al. have been used a Retina Net and proposed modified Retina Net for detecting and recognizing text from object images [28]. Heo et al. proposed a new worst-case execution time model using multipath neural network for DNNs on a GPU for object detection [29]. Qiang et al. recommended an object detection technique with joint segmentation approach [30].

9.3 Object Detection Techniques

9.3.1 Histogram of Oriented Gradients (HOG)

The HOG is a popular feature extraction method and it is widely used for object detection [31]. The following steps are used to calculate HOG feature descriptor.

- To reduce illumination effects, image normalization has to be applied to an image. In general, gamma correction is used for normalization of an image by calculating square root or log of an image. Finally, it reduces the illumination effects.
- In the next stage, calculate the first-order gradients using Gaussian smoothing with the discrete derivative masks.
- For calculating the gradient histograms, the image is split into sub-blocks and then find the 1D histogram of gradient of all the pixels in

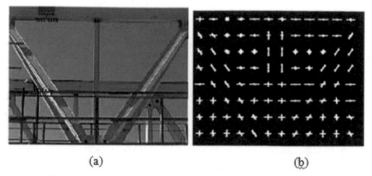

(a)   (b)

Figure 9.1   (a) Input image. (b) Features using HOG [31].

sub-block. Each histogram splits the gradient angle array into a fixed number of bins.

- Gradient strengths differ over a wide range due to local differences in lighting and foreground–background contrast, so to achieve good results, perform contrast normalize on each block and group all the blocks to form a large block.
- Finally, all the HOG descriptors converted into a combined feature vector. The Figure 9.1(a),(b) shows an input image and HOG feature descriptors.

9.3.2 Speeded-up Robust Features (SURF)

In many computer vision applications, feature extraction is an important task. The SURF algorithm is used for extracting the local features from images in order to identify the similarity between the images. The main advantage of this method is it computes very fast using box filters. This method also used for object detection and tracking in real-time applications [32]. The SURF mainly consists of interest point detection, interest point localization, interest point description, and fast indexing for matching. In the interest point detection blob features are extracted through performing the convolution between the integral image and determinant of the hessian matrix. The main aim of the interest point descriptor is to provide distinctive and strong feature descriptor of an image feature and it also invariant to rotation. To achieve invariant rotation, calculate the Haar-wavelet responses at which the key point was detected. The Figure 9.2 shows the detected strongest interest point in an image.

Figure 9.2   Interest points using SURF algorithm [32].

9.3.3 Local Binary Pattern (LBP)

LBP is widely used to extract texture features from an image. The LBP value of a pixel can be found out by thresholding the pixel values with the center pixel value. If the neighborhood pixel value is greater than the center pixel value, then that pixel is replaced with "1" otherwise replaced with "0." The binary digits are then multiplied with binomial values also called weighting function. Finally, the LBP value of that texture unit is obtained by summing the resulting pixel values [33]. Computation of 3×3 Local Binary Pattern (LBP) is shown in Figure 9.3.

9.3.4 Single Shot MultiBox Detector (SSD)

In real time applications, the mostly used object detection algorithm is single shot multibox detector (SSD) because of its easy implementation and good accuracy [34]. It is implemented using simple deep neural network. The SSD architecture is as shown in Figure 9.4. SSD increase the detection speed due to removal of redundant data.

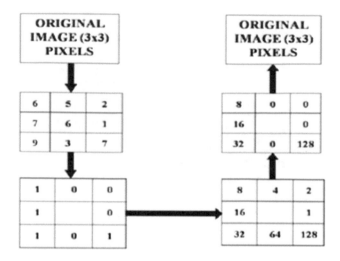

$$LBP = 8+16+32+128 = 184$$

Figure 9.3   Local Binary Pattern [33].

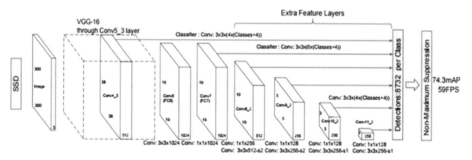

Figure 9.4   SSD Architecture [34].

The SSD object detection algorithm consists of two stages, one is feature extraction map and other one is applying convolutional filters to detect objects. In SSD algorithm, features of an image are extracted using Vgg16 predefined network. The Vgg16 network architecture is as shown in Figure 9.5.

It detects the objects using Conv4_3 layer. This algorithm creates four predictions from each cell. Each prediction gives 21 scores for each class. The highest scores bounding box is related to the bounded image of class. If many predictions do not have valid score, indicates that no object is detected.

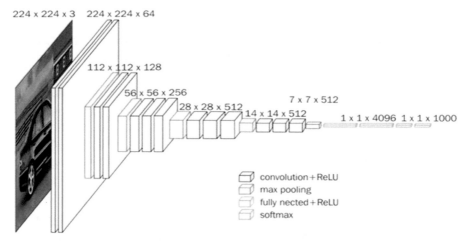

Figure 9.5 Vgg16 network architecture [34].

In the next stage, convolution filters are applied between the location and class scores to make the predictions. SSD algorithm can also be used for detecting multiple objects.

9.3.5 You Only Look Once (YOLO)

The YOLO algorithm is widely used for object detection and it is implemented using single DCNN. In YOLO method Google Net extract, the features of an objects obtained from the images [35]. The Google Net network architecture is as shown in Figure 9.6.

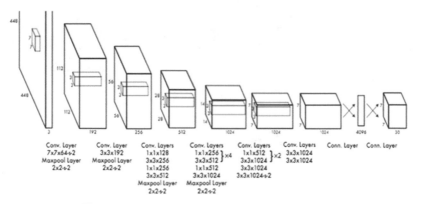

Figure 9.6 Google Net network architecture [35].

9.3.6 YOLOv1

YOLOv1 algorithm is fast compared to the conventional object detection methods, because it uses single deep convolutional neural network. It considers detection is a regression task but not a classification task [36]. In YOLOv1, the features are extracted from a full image and used to predict each bounding box. Since it only looks at the input image once, it predicts all bounding boxes in all classes within an image at the same time. In this algorithm, the input image is distributed into number of cells and it predicts the bounding boxes and object classification from each cell. It provides large number of bounding boxes, and finally those are consolidated for final prediction by a post processing procedure.

1. As mentioned in YOLO v1, each layer can only predict one class object. The grid cells predict two bounding boxes but a grid cell can only have one object in it. The presence of multiple objects results in shared space among bounding boxes with the different objects. This overlap causes confusion in the fully connected layer.
2. The model splits the input image to an S × S cell and each cell provides the bounding box predictions. Thus, due to the down sampling, the model uses rather coarse features to predict the bounding boxes.
3. It is difficult to localize objects or groups of minor objects. Hence, the main source of errors is localization.

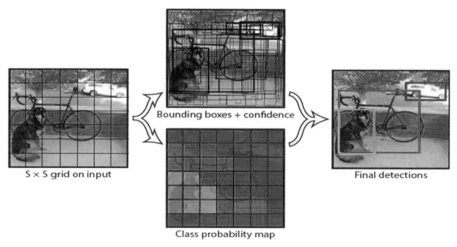

Figure 9.7   YOLOv1 System Model [36].

9.3.7 YOLOv2

YOLOv2 overcomes the limitations of YOLOv1 for object detection. YOLO gives low recall rate when compared with the region-based techniques. YOLOv2 mainly concentration on improving recall rate while retaining the classification accuracy. By adding the batch normalization in YOLOv1 mAP values are significantly improved [37]. With this modification, dropout layers are removed from the network without overfitting. The YOLOv1 suffers resolution problem at the output because it accepts the input image size 224 × 224 and convert to 448 for object detection. Hence, YOLOv2 tune the network with the image size 448 × 448 to overcome the resolution problem. Due to high resolution of tuning the network mAP increases up to 4%. YOLOv1 calculates the coordinates of bounding boxes with fully connected layers. It increases the complexity of object detection model. To reduce the complexity, YOLOv2 removes the fully connected layers from YOLOv1 and introduce anchor boxes for detecting the bounding boxes instead of coordinates. But object detection using anchor boxes reduces the mAP values and significantly increases the recall rate. The YOLOv2 architecture is as shown in Figure 9.8.

YOLOv2 uses K-means clustering algorithm to recognise the highest K bounding boxes from the images. The YOLOv1 does not have a control on location prediction, hence it makes unbalanced on initial iteration. YOLOv2 calculates five parameters and applies the sigma function to control

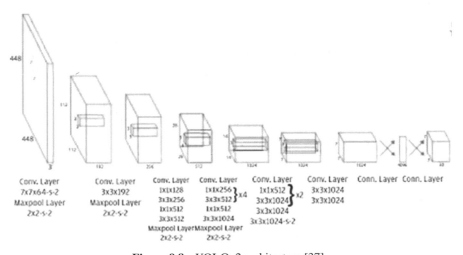

Figure 9.8 YOLOv2 architecture [37].

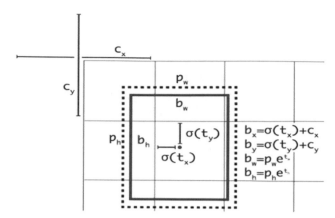

Figure 9.9   Bounding boxes with dimensions [37].

the values between 0 and 1. Boundary boxes with dimensions is as shown in Figure 9.9.

$$b_x = \sigma(t_x) + c_x$$
$$b_y = \sigma(t_y) + c_y$$
$$b_w = p w e^{tw}$$
$$b_h = p h e^{th}$$
$$p_r(object) * IOU(b, object) = \sigma(t_0)$$

where
$t_x, t_y, t_w,$ and t_h are predictions made by YOLO.
c_x and c_y are the top left corner of the grid cell of the anchor.
p_w and p_h are the width and height of the anchor.
$c_x, c_y, p_w,$ and p_h are normalized by the image width and height.
$b_x, b_y, b_w,$ and b_h are the predicted boundary box.
$\sigma(t_0)$ is the box confidence score.

YOLO v2 is trained on various predefined networks such as VGG-16 and GoogleNet. It is also trained on Darknet-19. Darknet network requires lower processing compared with other networks. The organization of Darknet-19 is given in Figure 9.11.

For detection of objects the last convolutional layer is replaced with three convolution layers of *1024 filters* followed by *1* × *1* convolution with the number of outputs.

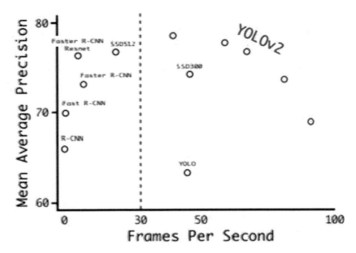

Figure 9.10   Accuracy and speed on VOC 2007 [37].

Type	Filters	Size/Stride	Output
Convolutional	32	3 × 3	224 × 224
Maxpool		2 × 2/2	112 × 112
Convolutional	64	3 × 3	112 × 112
Maxpool		2 × 2/2	56 × 56
Convolutional	128	3 × 3	56 × 56
Convolutional	64	1 × 1	56 × 56
Convolutional	128	3 × 3	56 × 56
Maxpool		2 × 2/2	28 × 28
Convolutional	256	3 × 3	28 × 28
Convolutional	128	1 × 1	28 × 28
Convolutional	256	3 × 3	28 × 28
Maxpool		2 × 2/2	14 × 14
Convolutional	512	3 × 3	14 × 14
Convolutional	256	1 × 1	14 × 14
Convolutional	512	3 × 3	14 × 14
Convolutional	256	1 × 1	14 × 14
Convolutional	512	3 × 3	14 × 14
Maxpool		2 × 2/2	7 × 7
Convolutional	1024	3 × 3	7 × 7
Convolutional	512	1 × 1	7 × 7
Convolutional	1024	3 × 3	7 × 7
Convolutional	512	1 × 1	7 × 7
Convolutional	1024	3 × 3	7 × 7
Convolutional	1000	1 × 1	7 × 7
Avgpool		Global	1000
Softmax			

Figure 9.11   Architecture of Darknet19 [37].

9.3.8 YOLOv3

YOLOv2 uses DarkNet-19 predefined network to extract the features, but YOLOv3 uses Darknet53 for feature extraction of an images. Darnet-53 consists of 53 convolutional layers and it is more efficient than Darnet-19 [38]. The DarkNet-53 network organization is given in Figure 9.12. In the YOLOv3 method, an input image is splits into a grid. The anchor boxes are calculated from each grid nearby objects. Each anchor box provides the score value, it gives the information about how accurately it predicts the object. One anchor box detects only one object ad gives one score value.

	Type	Filters	Size	Output
	Convolutional	32	3 × 3	256 × 256
	Convolutional	64	3 × 3 / 2	128 × 128
	Convolutional	32	1 × 1	
1×	Convolutional	64	3 × 3	
	Residual			128 × 128
	Convolutional	128	3 × 3 / 2	64 × 64
	Convolutional	64	1 × 1	
2×	Convolutional	128	3 × 3	
	Residual			64 × 64
	Convolutional	256	3 × 3 / 2	32 × 32
	Convolutional	128	1 × 1	
8×	Convolutional	256	3 × 3	
	Residual			32 × 32
	Convolutional	512	3 × 3 / 2	16 × 16
	Convolutional	256	1 × 1	
8×	Convolutional	512	3 × 3	
	Residual			16 × 16
	Convolutional	1024	3 × 3 / 2	8 × 8
	Convolutional	512	1 × 1	
4×	Convolutional	1024	3 × 3	
	Residual			8 × 8
	Avgpool		Global	
	Connected		1000	
	Softmax			

Figure 9.12   Darknet 53 architecture [38].

YOLOv3 is quick and accurate with respect to mAP and IoU values. YOLOv3 detects multiple objects by generating a multilabel for multiple bounding boxes. YOLOv2 detects single objects that is not suitable for some datasets like OID.

9.3.9 Regions with CNN (RCNN)

The simple and scalable algorithm for object detection is proposed by Girshick et al [39]. In RCNN algorithm, the input image features are extracted with convolutional neural network. In the first step, 2,000 region proposals are extracted using selective search algorithm. In the next step, 4,096-dimensional feature vectors from each region proposal using predefined network Alexnet were extracted, and all the features are fed to the Support Vector Machine (SVM) classifier to classify the object. The RCNN diagram, Alexnet architecture, and the object detection using RCNN is as shown in Figures 9.13, 9.14 and 9.15.

RCNN requires more time to classify the 2,000 region proposals per image. This is not suitable for real-time applications because it takes more time to train the network. Due to the selective search algorithm learning is not possible due to this bad region proposal are generated.

9.3.10 Fast RCNN

Fast RCNN overcomes the limitations of the RCNN to provide faster object detection algorithm [40]. The method is same as previous RCNN, only difference is instead of applying region proposal to convolutional neural network,

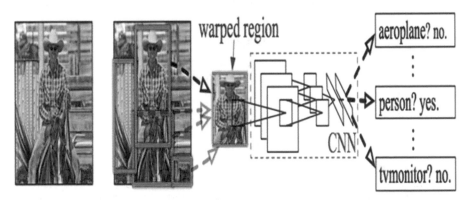

Figure 9.13 RCNN: Regions with RCNN [39].

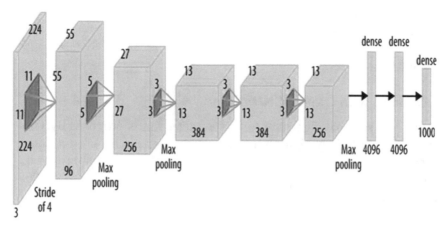

Figure 9.14   Architecture of Alexnet [39].

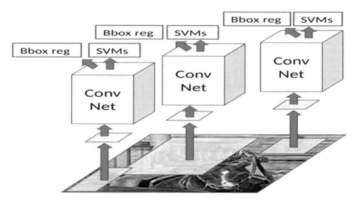

Figure 9.15   RCNN [39].

the CNN extract the features from the whole image to produce feature vector. The region proposals are identified from this feature vector and convert all into squares using pooling layer, then resize into fixed size, and then apply to fully connected layer. Finally, softmax layer is used to predict the class of region proposals and also offset values for the anchor box. Figure 9.16 shows the Fast RCNN method.

9.3.11 Faster RCNN

To extract the feature vector in Faster RCNN also like Fast RCNN, the input image is applied to CNN. In this algorithm, in place of selective

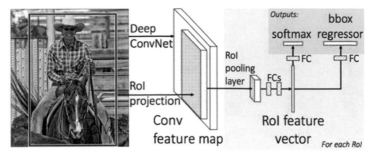

Figure 9.16 Fast RCNN [40].

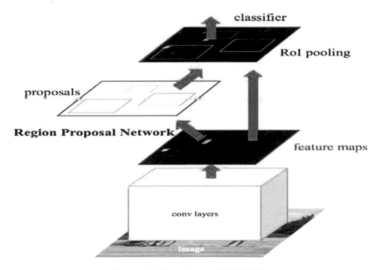

Figure 9.17 Faster RCNN [40].

search algorithm, a distinct network is used to identify the regions. The identified regions are the resized with pooling layer and next classify the image contained by the offered region and also calculate the offset values for the bounding boxes [41]. The Faster RCNN concept is as shown in Figure 9.17. Figure 9.18 shows the objects detected by Faster RCNN.

9.4 Datasets for Object Detection

Datasets have played an important role to analyze the algorithms. Most commonly used datasets for computer vision applications are given in Table 9.1. Figure 9.19 shows the objects detected on VOC 2007, VOC 2012,

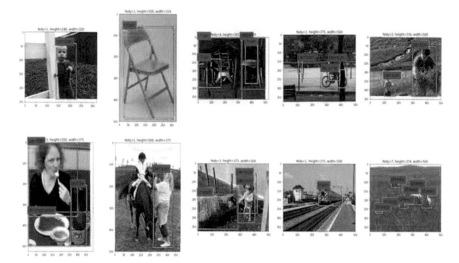

Figure 9.18   Various objects detected by Faster RCNN [40].

Table 9.1   Datasets used for Object Detection

Name of the Dataset	Number of Categories	Total Number of Images	Total Number Annotated Images
PASCAL VOC			
VOC 2005	4	1,282	2,209
VOC 2007	20	9,963	24,640
VOC 2009	20	7,054	17,218
VOC 2012	20	11,530	27,450
ILSVRC			
ILSVRC13	200	456,182	401,356
ILSVRC14	200	516,840	534,309
ILSVRC15	200	527,982	534,309
ILSVRC16	200	536,688	534,309
ILSVRC17	200	542,188	534,309
MS COCO			
MS COCO15	80	204,721	896,782
MS COCO16	80	204,721	896,782
MS COCO17	80	163,957	896,782
MS COCO18	80	163,957	896,782
Open images challenge object detection (OICOD)			
OICOD18	500	184,3041	11,535,515

and COCO datasets. Tables 9.3, 9.4 and 9.5 give the comparison of the different object detection methods including YOLO, YOLOv2, YOLOv3, RCNN, Fast RCNN, Faster RCNN, and SSD with mean average precision (mAP) values.

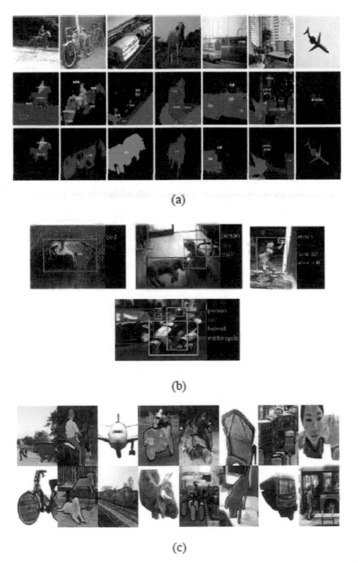

Figure 9.19 Objects Detection on (a) PASCAL VOC; (b) ILSVRC; and (c) MS COCO [40].

Table 9.2   Comparative Results on VOC 2007

Methods	Areo	Bike	Bird	Boat	Bottle	Bus	Car	Cat	Chair	Cow	Table	Dog	Horse	Mbile	Person	Plant	Sheep	Sofa	Train	Tv	mAP
RCNN [39]	68.1	72.8	56.8	43.0	36.8	66.3	74.2	67.6	34.4	63.5	54.5	61.2	69.1	68.6	58.7	33.4	62.9	51.1	62.5	68.6	58.5
RCNN [39]	73.4	77	63.4	45.4	44.6	75.1	78.1	79.8	40.5	73.7	62.2	79.4	78.1	73.1	64.2	35.66	66.8	67.2	70.4	71.1	66.0
FAST RCNN [40]	77.0	78.0	69.3	59.4	38.3	81.6	78.6	86.7	42.8	78.8	8.9	84.7	82.0	76.6	69.9	31.8	70.1	74.8	80.4	70.4	70.0
FASTER RCNN [41]	70	80.6	70.1	57.3	49.9	78.2	80.4	82.0	52.2	75.3	67.2	80.3	79.8	75.0	76.3	39.1	68.3	67.3	81.1	67.6	69.9
SSD300 [34]	80.9	86.3	79.0	76.2	57.6	87.3	88.2	88.6	60.5	85.4	76.7	87.5	89.2	84.5	81.4	55.0	81.9	81.5	85.9	78.9	79.6
SSD512 [34]	86.6	88.3	82.4	76.0	66.3	88.6	88.9	89.1	65.1	88.4	73.6	86.5	88.9	85.3	84.6	59.1	85.0	80.4	87.4	81.2	81.6

Table 9.3 Comparative Results on VOC 2012

Methods	Areo	Bike	Bird	Boat	Bottle	Bus	Car	Cat	Chair	Cow	Table	Dog	Horse	Mbile	Person	Plant	Sheep	Sofa	Train	Tv	mAP
RCNN [39]	71.8	65.8	52.0	34.1	32.6	59.6	60.0	69.8	27.6	52.0	41.7	69.6	61.3	68.3	57.8	29.6	57.8	40.9	59.3	54.1	53.3
RCNN [39]	79.6	72.7	61.9	41.2	41.9	65.9	66.4	84.6	38.5	67.2	46.7	82.0	74.8	76.0	65.2	35.6	65.4	54.2	67.4	60.3	62.4
FAST RCNN [40]	82.3	78.4	70.8	52.3	38.77	77.8	71.6	89.3	44.2	73.0	55.0	87.5	80.5	80.8	72.0	35.1	68.3	65.7	80.4	64.2	68.4
FASTER RCNN [41]	84.9	79.8	74.3	53.9	49.8	77.5	75.9	88.5	45.6	77.1	55.3	86.9	81.7	80.9	79.6	40.1	72.6	60.9	81.2	61.5	70.4
YOLO [36]	77.0	67.2	57.7	38.3	22.7	68.3	55.9	81.4	36.2	60.8	48.5	77.2	72.3	71.3	63.5	28.9	52.2	54.8	73.9	50.8	57.9
YOLOV2 [37]	88.8	87.0	77.8	64.9	51.8	85.2	79.3	93.1	64.4	81.4	70.2	91.3	88.1	87.2	81.0	57.7	78.1	71.0	88.5	76.8	78.2
SSD300 [34]	91.0	86.0	78.1	65.0	55.4	84.9	84.0	93.4	62.1	83.6	67.3	91.3	88.9	88.6	85.6	54.7	83.8	77.3	88.3	76.5	79.3
SSD512 [34]	91.4	88.6	82.6	71.4	63.1	87.4	88.1	93.9	66.9	86.6	66.3	92.0	91.7	90.8	88.5	60.9	87.0	75.4	90.2	80.4	82.2

Table 9.4 Comparative Results on MICROSOFT COCO

Methods	0.5:0.95	0.5	0.75	S	M	L	1	10	100	S	M	L
FAST RCNN [40]	20.5	39.9	19.4	4.1	20.0	35.8	21.3	29.4	30.1	7.3	32.1	52.0
FASTER RCNN [41]	24.2	45.3	23.3	7.7	26.4	37.1	23.8	34.0	34.6	12.0	38.5	54.4
YOLOV2 [37]	21.6	44.0	19.2	5.0	22.4	35.5	20.7	31.6	33.3	9.8	36.5	54.4
SSD300 [34]	23.2	41.2	23.4	5.3	23.2	39.6	22.5	33.2	35.3	9.6	37.6	56.5
SSD512 [34]	26.8	46.5	27.8	9.0	28.9	41.9	24.8	37.5	39.8	14.0	43.5	59.0

9.5 Conclusion

This Chapter discussed about various object detection methods for different applications under different conditions like occlusion, illumination effects on an image. It gives the information about handcrafted feature extraction methods including HOG, SURF, and LBP. It also provides the information about different versions of YOLO algorithm, RCNN, Fast RCNN, Faster RCNN, and SSD methods. Finally, this Chapter gives the comparison between the different algorithms on various datasets including VOC 2007, VOC 2012, and COCO. In future will develop efficient algorithms for multiobject detection and recognitions by reducing the test time.

References

[1] D. G. Lowe, "Object recognition from local scale-invariant features," *Proceedings of the Seventh IEEE International Conference on Computer Vision*, 1999, pp. 1150–1157 vol. 2.

[2] R. Fergus, P. Perona and A. Zisserman, "Object class recognition by unsupervised scale-invariant learning," *2003 IEEE Computer Society Conference on Computer Vision and Pattern Recognition, 2003. Proceedings.*, 2003, pp. II–II,

[3] D. G. Lowe, "Distinctive image features from scale-invariant key points," Int. J. of Comput. Vision, vol. 60, no. 2, pp. 91–110, 2004.

[4] K. Mikolajczyk and C. Schmid, "A performance evaluation of local descriptors," in *IEEE Transactions on Pattern Analysis and Machine Intelligence*, vol. 27, no. 10, pp. 1615–1630, Oct. 2005.

[5] R. Lienhart and J. Maydt, "An extended set of haar-like features for rapid object detection," in ICIP, 2002.

[6] C. Tang, Y. Feng, X. Yang, C. Zheng and Y. Zhou, "The Object Detection Based on Deep Learning," *2017 4th International Conference on Information Science and Control Engineering (ICISCE)*, 2017, pp. 723–728.

[7] L. Jiao *et al.*, "A Survey of Deep Learning-Based Object Detection," in *IEEE Access*, vol. 7, pp. 128837–128868, 2019.

[8] C. Cortes and V. Vapnik, "Support vector machine," Machine Learning, vol. 20, no. 3, pp. 273–297, 1995.

[9] Cover, T. M. and P. E. Hart. "Nearest neighbor pattern classification", IEEE Trans. on Information Theory IT-13: 21–27. 1967.

[10] Y. Freund and R. E. Schapire, "A desicion-theoretic generalization of on-line learning and an application to boosting," J. of Comput. & Sys. Sci., vol. 13, no. 5, pp. 663–671, 1997.

[11] Haigang Zhu, Xiaogang Chen, Weiqun Dai, Kun Fu, Qixiang Ye, Jianbin Jiao, "Orientation Robust Object Detection in Aerial Images using Deep Convolutional Neural Network", ICIP 2015, pp. 3735–3739.

[12] Anuj Mohan, Constantine Papageorgiou, and Tomaso Poggio, "Example-Based Object Detection in Images by Components", IEEE Transactions on Pattern Analysis and Machine Intelligence, Vol. 23, No. 4, April 2001, pp. 349–361.

[13] Zeng Z, Zhang J, Wang X, Chen Y, Zhu C. Place Recognition: An Overview of Vision Perspective. *Applied Sciences,* 8, no. 11: 2257, pp. 1–9.

[14] Josip Josifovski, Matthias Kerzel , Christoph Pregizer, Lukas Posniak, Stefan Wermter, "Object Detection and Pose Estimation based on Convolutional Neural Networks Trained with Synthetic Data", 2018 IEEE/RSJ International Conference on Intelligent Robots and Systems (IROS) Madrid, Spain, October 1–5, 2018, pp. 6269–6276.

[15] Mohana, HV Ravish Aradhya, "Object Detection and Tracking using Deep Learning and Artificial Intelligence for Video Surveillance Applications", International Journal of Advanced Computer Science and Applications, Vol. 10, No. 12, 2019, pp. 517–530.

[16] R. Girshick, J. Donahue, T. Darrell and J. Malik, "Region-Based Convolutional Networks for Accurate Object Detection and Segmentation," in *IEEE Transactions on Pattern Analysis and Machine Intelligence*, vol. 38, no. 1, 1 Jan. 2016, pp. 142–158.

[17] C. Wang and Z. Peng, "Design and Implementation of an Object Detection System Using Faster R-CNN," *2019 International Conference on Robots & Intelligent System (ICRIS)*, 2019, pp. 204–206.

[18] X. Zhou, W. Gong, W. Fu and F. Du, "Application of deep learning in object detection," *2017 IEEE/ACIS 16th International Conference on Computer and Information Science (ICIS)*, 2017, pp. 631–634.

[19] K. L. Masita, A. N. Hasan and T. Shongwe, "Deep Learning in Object Detection: a Review," *2020 International Conference on Artificial Intelligence, Big Data, Computing and Data Communication Systems (icABCD)*, 2020, pp. 1–11.

[20] Q. Zhang, N. Huang, L. Yao, D. Zhang, C. Shan and J. Han, "RGB-T Salient Object Detection via Fusing Multi-Level CNN Features," in *IEEE Transactions on Image Processing*, vol. 29, 2020, pp. 3321–3335.

[21] Asim Suhail, Manoj Jayabalan, Vinesh Thiruchelvam. CONVOLU-
TIONAL NEURAL NETWORK BASED OBJECT DETECTION: A
REVIEW. JCR. 2020; 7(11): 786–792.

[22] Tanvir Ahmad, 1 Yinglong Ma, 1 Muhammad Yahya, 2 Belal Ahmad,
3 Shah Nazir, 4 and Amin ulHaq, "Object Detection through Modified
YOLO Neural Network", Article in Scientific Programming, June 2020,
pp. 1–10.

[23] Gu J, Lan C, Chen W, Han H. Joint Pedestrian and Body Part Detection
via Semantic Relationship Learning. *Applied Sciences*. 2019; 9(4):752,
pp. 1–14.

[24] Liu, L., Ouyang, W., Wang, X. *et al*. Deep Learning for Generic Object
Detection: A Survey. *Int J Comput Vis* **128,** 261–318 (2020).

[25] Christian Szegedy, Alexander Toshev, DumitruErhan, "Deep neural networks for object detection", NIPS'13: Proceedings of the 26th International Conference on Neural Information Processing Systems - Volume
2, December 2013, pp. 2553–2561.

[26] Z. Zhao, P. Zheng, S. Xu and X. Wu, "Object Detection With Deep
Learning: A Review," in *IEEE Transactions on Neural Networks and
Learning Systems*, vol. 30, no. 11, pp. 3212–3232, Nov. 2019.

[27] Wang EK, Li Y, Nie Z, Yu J, Liang Z, Zhang X, Yiu SM. Deep Fusion
Feature Based Object Detection Method for High Resolution Optical
Remote Sensing Images. *Applied Sciences*. 2019; 9(6):1130.

[28] Siba Haidar1*, Ihab Sbeity1 and MarwaAyyoub, "Text Detection using
Object Recognition Techniques", COJ Robotics & Artificial Intelligence, vol 1, issue 1, 2019, pp. 1–6.

[29] S. Heo, S. Cho, Y. Kim and H. Kim, "Real-Time Object Detection
System with Multi-Path Neural Networks," *2020 IEEE Real-Time and
Embedded Technology and Applications Symposium (RTAS)*, 2020,
pp. 174–187.

[30] Qiang B, Chen R, Zhou M, Pang Y, Zhai Y, Yang M. Convolutional Neural Networks-Based Object Detection Algorithm by Jointing Semantic
Segmentation for Images. *Sensors*. 2020; 20(18):5080.

[31] N. Dalal and B. Triggs, "Histograms of oriented gradients for human
detection," in CVPR, 2005.

[32] Bay, H., A. Ess, T. Tuytelaars, and L. Van Gool. "SURF: Speeded Up
Robust Features." *Computer Vision and Image Understanding (CVIU)*.
Vol. 110, No. 3, pp. 346–359, 2008.

[33] S. R. Dubey, S. K. Singh and R. K. Singh, "Multichannel Decoded
Local Binary Patterns for Content-Based Image Retrieval," in *IEEE*

Transactions on Image Processing, vol. 25, no. 9, pp. 4018–4032, Sept. 2016.

[34] W. Liu, D. Anguelov, D. Erhan, C. Szegedy, S. Reed, C. Fu, and A. Berg. *ECCV (1), Lecture Notes in Computer Science, volume 9905* , pp. 21–37, Springer, 2016.

[35] J. Redmon, S. Divvala, R. Girshick, and A. Farhadi, "You only look once: Unified, real-time object detection," in Proceedings of the Conference on Computer Vision and Pattern Recognition, pp. 779–788, CVPR Press, July 2016.

[36] Juan Du, "Understanding of Object Detection Based on CNN Family and YOLO", IOP Conf. Series: Journal of Physics: Conf. Series 1004, 2018, pp. 1–8.

[37] J. Redmon and A. Farhadi, "YOLO9000: Better, Faster, Stronger," *2017 IEEE Conference on Computer Vision and Pattern Recognition (CVPR)*, 2017, pp. 6517–6525.

[38] Redmon, Joseph &Farhadi, Ali. (2018). YOLOv3: An Incremental Improvement.

[39] R. Girshick, J. Donahue, T. Darrell, and J. Malik, "Rich feature hierarchies for accurate object detection and semantic segmentation," in CVPR, 2014.

[40] R. Girshick, "Fast r-cnn," in ICCV, 2015.

[41] S. Ren, K. He, R. Girshick, and J. Sun, "Faster r-cnn: Towards real-time object detection with region proposal networks," in NIPS, 2015, pp. 91–99.

10

An Intelligent Patient Health Monitoring System Based on a Multi-scale Convolutional Neural Network (MCCN) and Raspberry Pi

Shruti Bhargava Choubey and Abhishek Choubey

Department of Electronics Communication, Sreenidhi Institute of science and technology, Hyderabad, India
E-mail: shrutibhargava@sreenidhi.edu.in; abhishek@sreenidhi.edu.in

Abstract

An intensive care system is a structure or a device that can support family members in monitoring patients the slightest bit. It is capable of providing security and comfort to the family members. Lately, smart wearable's, characteristically as the wearable electrocardiogram (ECG), electromyography (EMG), electroencephalography (EEG), photoplethysmography (PPG), blood pressure (BP), heart sound, respiration, sleep, and motion monitoring, looking to be an immense and talented market in the expertise industry. Those clinically and scientifically valuable for healthier intensive care of real-time, long-term, and dynamic pathological and physiological expansions, thus, so long as occasions for the progress of innovative diagnostic and therapeutic procedures.

The idea of machine learning (ML) involves the exploitation of hardware gadgets that screen or catch information and are related to the open or private cloud, empowering them to naturally trigger certain occasions. In this chapter, significant emphasis has been given to an ML for an efficient real-time monitoring system. The basic purpose of the chapter is to explore the numerous possibilities of ML in the field of real-time monitoring systems.

Keywords: Signal, Raspberry Pi, MCCN.

10.1 Introduction to Signal Processing

An embedded system is a particular reason structure where the PC is completely embodied through or devoted to the machine or structure it controls. By no means like an all-around valuable personal computer (PC), for example, has a PC, an embedded system accuracy single or a small number of predefined errands, customarily with focused necessities. Because of the situation that the structure is devoted to remarkable errands, plan fashioners can build it, decreasing the measurements and cost of the thing.

Settled in traces are customarily group-conveyed, benefitting from first to last business sectors of range. Personality developed partners (PDAs) or handheld PCs are recurrently intention to be embedded effects because of the intention of their wire's setup, in spite of the information that they are additional stretchy in indoctrination phrases [43].

This column of explanation keeps up on clouding as substance creates (1) by the preface of the OQO show and (2) by means of the dwelling windows XP operational construction and ports, in favor of case, a USB port, doubles the features when in stipulation have a vicinity with "incredibly profitable PCs." These procession of classifying clouds hugely supplementary inserted systems expect real part in tools changes as of smallest diplomacy to goliath still organizations similar to propelled watches and an MP3 group of actors, weight group illumination, fabricating plant controllers, or the system scheming nuclear life vegetation. So far as multifaceted design embedded systems can keep running from especially ordinary with a private microcontroller break off in the direction of absolutely complex with endless models, peripherals, and projects mounted within a broad body otherwise fenced in the discipline.

The data from sensors are collected from the serial port and analyzed with classification using MATLAB. MATLAB is a powerful tool for engineering and scientific languages. For the development of MCNN, in MATLAB, almost all MCNN toolboxes are available, and with the help of available toolbox, we can achieve better performance.

CNNs use a distinction of multilayer perceptions intended to require negligible preprocessing. They are likewise recognized as shift invariant or space invariant artificial neural networks (SIANN), grounded on their shared-weights architecture and translation invariance features [29, 30].

An implanted framework is a distinct cause framework where the computer is utterly epitomized via or dedicated to the machine or framework its wheel. Under no circumstances similar to a universally helpful laptop, intended pro instance, a laptop, an inserted framework performs one or a pair of natural everyday jobs, ordinarily with targeted necessities. Due to the circumstance that the framework is dedicated to unique errands, plan designers be able to increase them, reducing the dimensions and price of the item. Entrenched outlines are traditionally mass-delivered, profiting through markets of size.

10.1.1 Cases of Implanted Frameworks

- Cellular telephones and mobile switches.
- Handheld adding together machines.
- Handheld PCs.
- Family apparatuses, together with microwave broilers, garments washers, TVs, DVD avid gamers, and recorders.
- Scientific hardware.
- Private computerized correct hand.
- Videogame supports.
- Laptop peripherals, for instance, switches and printers.
- Industrial controllers for remote computing device venture.

An implanted framework is being whose PC gear with brainwashing installed in it as considered one of its ingredients. We are able to state that it is "A blend of microprocessor equipment and training, and maybe extra perfunctory or one-of-a-kind mechanism, intended to play out a faithful possible. Repeatedly, ingrained context is a part of a higher framework or thing, the same to the example of stopping automation in an auto." The architecture of embedded system is shown in Figure 10.1 An installed framework is a distinguished underlying principle computer framework planned to play out unique dedicated capacity. It is mainly set up as a primary aspect of a complete machine including equipment and mechanical components.

A well-known framework is a combination of indoctrination and equipment to play out a faithful errand. A section of the major objects utilized as a piece of installed items are microchips and microcontrollers

Centrality: Due to the statistic that of their diminished measurement, low rate, and easy map highlights made entrenched tasks highly extensive and infringed into human lives and have boost to be imperative. They are observed in every state of affairs from kitchen manufactured goods to condominium make.

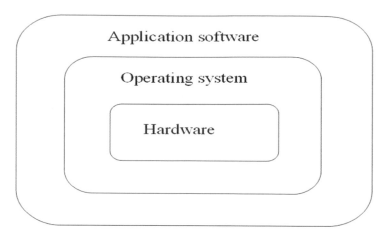

Figure 10.1   Architecture of an embedded system.

Every implanted framework comprises custom-manufactured gear worked around a focal making ready unit (CPU). This gear additionally includes reminiscence chips against which the product is stacked. The product living on the reminiscence chip is additionally referred to as the "firmware." The working procedure keeps jogging over the gear, and the applying programming continues strolling over the working framework. The indistinguishable establishment applies to any computer including a computer.

Be that as it is going to, there are large contrasts. It is not obligatory to have a working process in every inserted for little home gear, for instance, remote manage units, aeration and cooling systems, toys and countless others, there is not a want for an operational method and we will compose satisfactory the application targeted to that utility.

For purposes together with dubious handling, that you just will have to have a working procedure. In any such case, you ought to consolidate the apparatus programming with the working manner after which trade the entire programming onto the reminiscence chip. When the applying is exchanged to the memory chip, the application will continue to hold strolling for quite an even as refill story claim.

10.1.2 Features of Embedded Systems

Implanted frameworks complete a separate errand they cannot manage to survive customized to do distinctive belongings. Set up frameworks contain primarily constrained material goods, positively this memory. When all is claimed in finished, they do not need supplementary stockpiling contraption such in mild of the reality that the CDROM or the floppy plate. Set up frameworks require work contrary to a few deadlines.

An instantly recognizable recreation should be informed inside a detailed time. In some entrenched frameworks, known as regular systems, the focuses in time are severe. Lacking a futile line may with ease reason a disaster—lack of presence or injury to property. Inserted frameworks are compelled for manipulating, the equal quantity of implanted frameworks work via a battery, the vigor operation needs to below.

Inserted frameworks must be intensely strong. Each so as a rule, squeezing ALT-CTRL-DEL is very well to your work discipline, nevertheless you cannot stand to reset your implanted framework.

Some inserted systems have got to work in severe normal stipulations, for instance, primarily excessive temperatures and humidity. Established packages complete and totally point by means of factor errands, they cannot be modified to do selective problems. Implanted tasks that control the supporter promote (for illustration computerized toys) are totally expense-powerful. Indeed [44, 45] and [46], even a rebate of Rs.10 is a part of price sparing, considering the statistic that incalculable numbers or thousands of frameworks would likewise be offered.

Dislike registering system work areas where the apparatus stage is ruled through Intel and the working technique is dominated through Microsoft; there may be a massive variety of processors and dealing tasks for the implanted projects. In this means, determining a determination on the correct stage is truly probably the most multifaceted venture.

10.1.3 Domain Applications

The conversation server shops the arriving statistics of numerous topics in a software-based database and makes it possible for entry from in every single place the sector through an internet-based administration process. The developed method enables one inspector to watch the biological limitations and the particular capacity of a few subjects in a one-of-a-kind place [25, 49, 50]. Thereby the themes can transfer practically easily in any extent blanketed by the mobile community and extra security to the field as well as to the examiner for an evaluation. An online-established administration request enables a constant far flung monitoring of the physiological and the subjective knowledge of the area.

10.2 Background of the Medical Signal Processing

10.2.1 Literature Review

Plenty of instant arrangement classification calculations projected. Most customary TSC techniques fall into two categorizations: remove put-together strategies to facilitate utilization ken classier though respect to best of separation measures between instant arrangement and highlight based classifiers to facilitate concentrate or look for deterministic highlights in clock recurrence space and after to facilitate apply conventional sorting procedures. In current centuries, a few collaborative approaches to facilitate accumulate many TSC classier collected have additionally been considered. A complete assessment of these techniques is out of choice right here, but we can do a wide-ranging experiential evaluation through leadings methods in the next segment. Below, we evaluation some works which are most associated though McNeil latest years, there were lively studies on deep neural networks [1, 4, 13, 25, 43] to facilitate many associations graded characteristic abstraction and arrangement together. Extensive contrast has been shown to facilitate complication procedures in CNN have a better capability on extracting meaningful features than Adhoc attribute choice [21]. However, programs of CNN children have now not been studied until these days. A multi-channel CNN has been proposed to deal through multivariate instant series [32].

Features are extracted by utilizing settings on every occasion collection into different CNNs. From to facilitate point forward, they connect those highlights and place them into another CNN system. Extensive multivariate datasets are required to prepare this profound engineering. While for our technique, we center approximately univariate instant arrangement and present two additional branches to facilitate can extricate multiscale and multirecurrence data, and further increment forecast precision.

Encourages CNN though factors [8, 43] posthandled utilizing an info variable choice (IVS) calculation. Key deference compared through multi-scale convolution neural network is to facilitate as they go for diminishing info estimate through different IVS calculations. Conversely, we are investigating increasingly crude data for CNN to find. Notwithstanding classification, CNN is additionally utilized for instant arrangement metric learning. In [33], Zhen et al. proposed a model called convolution nonlinear neighborhood components examination to facilitate performs CNN based measurement learning and uses 1-NN as more tasteful in implanting space.

Author in [29] aiming at the existing issues of an electronic mobile healthcare system a distinct framework HES is proposed. The implementation of an expert system that primarily addresses routine physical examinations can greatly reduce a doctor's or administrator's involvement and enable families and guardians to access users' health information anytime and anywhere. Therefore, HES can serve as a significant component of the Informa ionization of medical industries. However, some problems remain unsolved, for example, the diagnosis reliability of the expert system is not perfect, and HES cannot currently monitor or analyze sudden diseases.

In [17], there are various parameters in this system stated that ECG, heart rate, variability, pulse oximetry, plethysmography, and fall detection. The tele-medical system focuses on the measurement and makes an evaluation of these parameters. In android phones, there are two different designs of wireless body networks. Data acquisition in the (W) ban plus the use of the smartphone sensors, data transmission, and emergency communication is connected to an embedded system.

With first responders and clinical server, it is important to smart and energy-efficient sensors. In Zigbee-based approach, sensor nodes obtain physiological parameters and perform signal processing, data analysis, and transmit measurement value to the node. In the second design through cable, sensors are connected to an embedded system. In this system, Bluetooth is used for transferring data to Android.

Authors in [27] stated that as we are dealing with e-healthcare monitoring system, our system is based on a wireless sensor network (WSN) and smart devices. It is important to have a network between the doctor, patient, and caretakers to monitor the condition of the patient. In this, sensors are used for monitoring the patient health parameters in surroundings. Sensors are passed to prior devices through transmitter and end-user. In this system, doctors and caregivers can observe the patient's health condition without visiting. And they can also update the medication and upload the patient's medical reports on the Web server and later can be accessed by the patient anywhere. It is an easy process and very convenient for both doctors and patients. With the help of this data, doctors can observe and understand patient conditions from private home patient to public healthcare center patient. This technique is cost reducing.

Author in [26] stated that the Android application called "ECG Android App" is developed for the healthcare system which is based on IoT and Cloud, and it provides the end user with visualization of their electro cardiogram (ECG) waves and data logging functionality in the background. The logged data can be uploaded to the users private centralized cloud that can be monitored by the patient as well as the doctor.

10.2.2 Problem Identification

I. Various research and effort has long gone into the making of a better and well-increased healthcare monitor method, even for distantly placed patients. Only some of those researches have considered environmental explanations to facilitate have an impact on human wellness state [39, 40]. Thus, this chapter places forth need, process, and more than a few tendencies in utilizing both scientific and environmental sensors. Moreover, taking potential of developed technological know-how available to most people, a far-off healthcare monitoring approach that is ready to supply wellbeing parameters though the aid of Wi-Fi approach to smartphones of every sufferer and health practitioner; it is possible to furnish instant prognosis with no trouble by clicking a button.

II. It is distinct to facilitate the application of current applied sciences surrounded by an area of patient monitoring that has enormous advantages for communities [36, 37]. Pick an important person of those technologies in a country-wide stage must be recognized on better working out of those tools, examine their advantages and downsides, and based on conditions, technical, human, and fiscal resources in healthcare corporations, affordability of these amenities, strategic wants and challenge to take the position.

III. It is obvious to facilitate the suitable application of every one of those methods in the nation is offered though determine and get to the bottom of technical and nontechnical challenges that are taking part in a crucial position surrounded by the triumphant implementation of these applied sciences; comprehend benefits of those technologies. Triumphant implementation of mobile-based programs is always faced through challenges corresponding to increasing accuracy of critical alerts, interoperability between one-of-a-kind programs, bandwidth obstacles, best of health offerings, battery lifestyles restrained tools, and so forth. The summary of previous work is shown in Table 10.1.

10.3 Real-Time Monitoring Device

10.3.1 Hardware Design Approach

Every part read and utilized the educated assent subsequent to having the opportunity to cooperate with the examination hardware and make inquiries. Members were given contact data to the examination nurture who might catch

Table 10.1 Summary of research work.

Title	Methodology Used	Issues
Secure Management of Personal Health Records by Applying Attribute-Based Encryption [9]	A novel modification of the CPABE system is used. Data storage and admittance control mechanism is used.	No formal security proof
Effective Ways to Use the Internet of Things in the Field of Medical and Smart Healthcare [16]	m-Health and e-Health are providing diverse services in the least, such as preclusion and diagnosis counter to disease, risk calculation, intensive care patient health, tutoring, and handling to users	Only data storing, no access control schemes
Internet of Things: Remote Patient Monitoring Using Web Services and Cloud Computing [11]	Electro Cardiogram (ECG) Waves and Data Logging Functionality	No field test was conducted
The New Secure and Efficient Data Storage Approaches for Wireless Body Area Networks [6]	the ABE-based contact control technique is extra talented than other procedures of attaining all the safe-keeping necessities.	Not addressing the security-safety conflict.
Private and Secured Medical Data Transmission and Analysis for Wireless Sensing Healthcare System [4]	The prototype implementation of HES is explored to verify its feasibility.	The diagnosis reliability of the expert system is not perfect, and HES cannot currently monitor or analyze sudden diseases

up with them to plan their first in-home nursing visit. Much of the time, the remote network implied that the establishment and preparation, as a rule, took under 1 h [25]. Members were given the TEQ and SUS questions and asked to far-reaching and profit the medical attendant for their first visit. This was done to guarantee that candidates had some involvement with the framework before rating ease of use. The SUS has managed again when the framework was expelled a half year later. A design approach is shown in Figure 10.2.

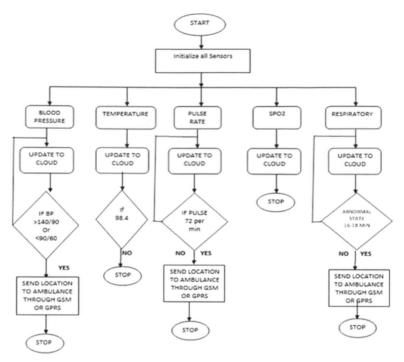

Figure 10.2   A design approaches.

10.3.2 Multi-Scale Convolutional Neural Networks

Time series classification has become a challenging problem. In this chapter, we are proposing a MCNN framework for the TSC. Here, we are giving input as predicted time series and output is its label. In the MCNN, totally there are three stages to calculate the TSC. The first stage is the transformation stage, the second stage is the local convolution stage, and the third stage is the full convolution stage. The block diagram of proposed method is shown in Figure 10.3. In the first stage, we are applying identity mapping, down-sampling transformations in the time domain spectral transformations in the frequency domain to the input time series. In the second stage, we are extracting features for every branch and with different sizes, outputs are passes through the max-pooling process. In the final stage, we are combining all the features and applying some more convolutional layers for the final output. Using all the steps and backpropagation algorithms, we are calculating features in frequency, time scales, and time series. The major applications are biomedical engineering and clinical applications.

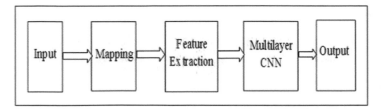

Figure 10.3   Block diagram of proposed method.

For real-time applications, the TSC will change every time. For example, for different persons, the temperature, systolic, diastolic, heartbeat, and Spo2 are going to be different. So, every time it should be updated using the neural networks, we are able to update the TSC. Using the proposed method, we are calculating all these parameters and updating depends on the requirement.

10.3.3 Raspberry Pi

The Raspberry Pi 3 exhibit B has especially worked with the Broadcom BCM2837 System-On-Chip (SoC) joins four prevalent ARM Cortex-A53 process focuses running at 1.2 GHz with 32 Kb Level one and 512 Kb Levels 2 or 3 hold memory, a Video Core IV representations processor, and is related with a 1-GB LPDDR2 memory module on the back of the board. According to the affiliation astute, the board should be prepared for sending data to and from the board rapidly. Another twofold band Wi-Fi supports for 2.4 and 5 GHz 802.11b/g/n/cooling which similarly ensures twofold all through the 802.11b/g/n/cooling Wi-Fi on the Raspberry Pi 3 Model B. With the development of Gigabit Ethernet over USB 2.0, the wired Ethernet execution is upheld, with a phenomenal throughput of around 300 Mb.

10.3.4 16×2 Liquid Crystal Display (LCD)

The fluid precious stone show might be an extremely real contraption in the inserted method. Nowadays, it is essentially standard for uncover industry to make utilization of LCD supplanting Cathode Ray Tubes (CRT). Pixels are utilized for some bendy ones.

10.3.5 Ubidots

In this intractable, we will show how to use the Ubidots platform to publish data coming from an interface developed in LabVIEW. The get and post

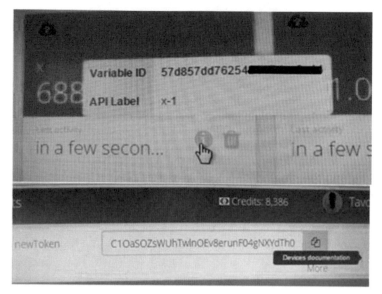

Figure 10.4   Variable ID is a unique alphanumeric and verification ID.

methods can be used to link the virtual instrument with the Ubidots variables. Figure 10.4 shows the G code to perform the connection with the web service to post data.

10.3.6 Blood Pressure Module

Circulatory strain and pulse examining are demonstrated in plain view with serial out for outside activities of inserted circuit handling and show. Shows systolic, diastolic, and pulse readings. A smaller plan coordinates over your wrist like a watch. Simple to utilize wrist shape wipes out pumping.

10.3.7 Temperature Sensor (TMP103)

A thermistor is a kind of resistor with the opposition is elegant on temperature. Thermistors are typically utilized as inrush current limiter, temperature sensors (NTC style for the most part), self-resetting over remuneration defenders, and automatic warming components. The TMP103 advanced yield temperature sensor in a four-ball wafer chip-scale package (WCSP). The TMP103 handles considering temperature to a goal of 1°C. In Figure 10.5 shows project variables showing the most recent acquired values

Figure 10.5 Project variables showing the most recent acquired values.

10.3.8 Respiratory Devices

Once regulatory or triggering high-quality respirational plans, an energetic feature is the initial recognition of the patient's inhalation stage over a movement trigger. This is the one method by which the expedient can assistance aim pulsive sniff with a preset overpressure and at a similar time retain the patient's respiratory energy to a smallest. In accumulation to this, the capacities should be extremely precise across the whole flow variety for numerous behaviors, so that the patient's respiratory outline can be perceived dependably. In present respiratory plans, an extremely dynamic discrepancy bulk sensor or a highly sensitive thermal mass flow sensor often screens the whole respiratory motion and the spur-of-the-moment breathing effort of the patient.

10.3.9 Updation of Data Using MCNN and MATLAB

This classifier is very effective for real-time data analysis. MCNN is the type of CNN. It is very efficient compared to state of art methods. The sensors like temperature sensor, pulse sensor, BP serial module, Spo2 sensor, heartbeat sensor, respiratory sensor are checked and updated for their linear values using MATLAB software in which there are built-in functions to develop it. The values for systolic, diastolic, pulse rate, oxygen saturation level, temperature, and respiratory are updated by feedback MCCN.

10.4 Outcome and Discussion

These arrangement settings are whole in addition to procedure comes to online and liquid crystal display show translates to "IOT wellness Care."

Temperature is calculated in addition to displayed resting on liquid crystal display show have "Temperature No DEG" place does not show related price. Subsequent stair for synchronizing spirit fee along though it is presented happening liquid crystal display has "Sync Heartrate.". After that rhythm depends on the support of 15 s with an indication on LCD shows "PULSE: NO" during initial column with complete considered spirit cost during second column "Heartrate: No BPM" does not show calculated worth. Figures 10.6 and 10.7 shows systolic blood pressure graph in ubidots and diastolic blood pressure graph in ubidots respectvely.

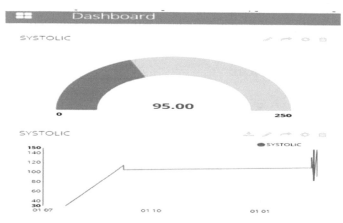

Figure 10.6   Systolic blood pressure Graph in ubidots.

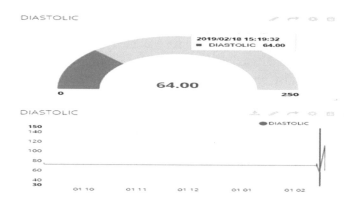

Figure 10.7   Diastolic blood pressure Graph in ubidots.

The above graph describes the systolic blood pressure in ubidots and updated in the cloud when the systolic value range is 95. When the systolic value exceeds more than 140(>140) or lesser than 90(<90) means the patient is in a dangerous condition. If the patient is in the dangerous condition, the message is sent to the doctor and the ambulance with the location using GSM and GPRS.

Along with the diastolic blood pressure in ubidots and updated in the cloud when the diastolic value range is 95. When the diastolic value exceeds

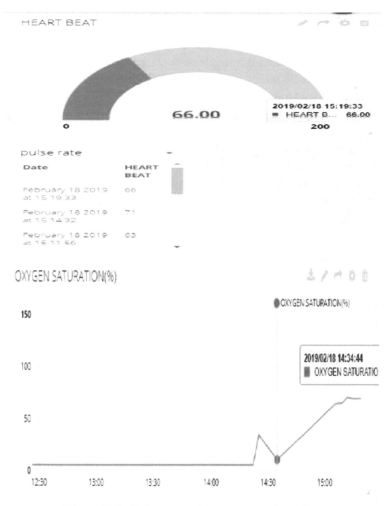

Figure 10.8   Pulse rate and oxygen saturation value.

more than 100(>100) or lesser than 60(<60) means the patient is in a dangerous condition. If the patient is in a dangerous condition, the message is sent to the doctor and the ambulance with the location using GSM and GPRS.

The above describes the heartbeat value (i.e., pulse rate) in ubidots and updated in the cloud when the pulse rate value range is 66, and the oxygen saturation range (i.e., SPO2) in ubidots and updated in the cloud when the Spo2 value range is 90. If the patient is in a dangerous condition, the message is sent to the doctor and the ambulance with the location using GSM and GPRS.

The above describes the temperature value in ubidots and updated in the cloud when the temperature value range is 28. Experimental research shows that the respiratory value at the normal case and the emergency case in ubidots and updated in the cloud when the respiratory value range is in normal case. The above describes the temperature value in ubidots and updated in the cloud when the temperature value range is 27 and the heartbeat value (i.e., pulse rate) in ubidots and updated in the cloud when the pulse rate value range is 66. The oxygen saturation range (i.e., Spo2) in ubidots and updated in the cloud when the Spo2 value range is 63. The real-time value of pulse rate and oxygen saturation is shwon in Figure 10.8.

In the systolic blood pressure in Matlab when the systolic value range is 102 and diastolic blood pressure in ubidots and updated in the cloud when the diastolic value range is 69.

10.5 Conclusion

Among this extensive utilization of the web, these efforts be concentrated in the direction of put into effect this web knowledge headed for setting up a structure would during this technology on behalf of superior physical condition. The technology of matters is anticipated to regulate the sector within additional than a few fields except extra advantage could subsist here the subject of physical condition. Accordingly, this effort is finished in the direction of intending an IoT headquartered neat health condition procedure via an ARM microcontroller. Within this effort, the MCP6004 founded pound oximeter be premeditated in addition to DS1820B temperature sensor be used in the direction of examining the temperature and coronary sensitivity rate of the sufferer and the microcontroller pick up the info as well as ship this via GPRS protocol.

The info is also dispatched to the LCD for the show so the sufferer can understand his wellness reputation. In the path of the intense situation to prepare the medical professional forewarning, communication is sent to the healthcare professional's phone by means of GSM modem connected and even as the buzzer turns to aware the porter. The medical professionals can view the sent data by classification to the HTML webpage making use of specified IP and page fresh alternative is prearranged, and as a result, constant information reaction can be achieved. As a result, a steady patient monitoring approach is designed.

We have exhibited a MCNN, a convolution neural system customized foretime arrangement classification. MCNN joins highlight extraction and classification, and mutually learns parameters through back proliferation. It uses the quality of CNN to naturally adapt great element portrayals in both instant and recurrence spaces. Specifically, MCNN contains numerous branches to facilitate and perform dissimilar changes of instant arrangement, which extricate highlights of different recurrence and instant scales, tending *to restriction of numerous past works to facilitate they just concentrate* highlights at a solitary instant scale. We have additionally discussed insights to facilitate studying convolution alters in MCNN generalizes shape let mastering, which imparts explains the extraordinary overall performance of MCNN.

More importantly, a bonus of CNNs is they can fascinate a massive number of facts to analyze accurate function illustrations. Right now, all TSC datasets we approach are not extremely expansive, going from a preparation size of approximately 50 to a couple of thousands.

We imagine facilitating a MCNN will demonstrate even greater preferences later on when prepared through a lot bigger datasets. We trust MCNN will move more research on incorporating profound learning through instant arrangement information examination. For future work, we will explore how to increase Mannford instant arrangement classification by consolidating opposite side data from dissimilar sources, for example, content, pictures, and discourse.

10.6 Future Work

To be able to put in force future improvements to the wellbeing monitoring network, we are able to introduce new sensors such as well as location monitoring capabilities. We can also plan to integrate alarm triggering algorithms and developed security systems in wireless sensor networks which would be predominant in a wellness monitoring atmosphere.

It presents continuous monitoring of the important signs of the sufferer over long durations of time unless an irregular situation is captured and hence valuable circumstances may also be overcome. This wellness monitoring method presents long-run monitoring capacity priceless for the staff in the hospitals and reduces their workload.

Future work may just incorporate an extra number of sensors in a single procedure to provide flexibility. An additional facet to remember is involving wearable sensors or wearable sensors that can well be more cost-effective and more diverse in utility which will help in improving the already current systems.

Acknowledgments

The authors would like to thank the Sreenidhi Institute of Science and Technology, Hyderabad for providing the infrastructure to conduct the research.

References

[1] P. S. Teh, A. B. J. Teoh, and S. Yue, "A survey of keystroke dynamics biometrics," The Scientific World Journal, vol. 2013.

[2] T. Sim, S. Zhang, R. Janaki Raman, and S. Kumar, "continuous verification making use of multimodal biometrics," IEEE Trans. Sample analysis and computer Intelligence, vol. 29, no. Four, pp. 687–seven-hundred, 2007.

[3] J. Bonneau, C. Herley, P. C. Van Orschot, and F. Stajano, "The hunt to exchange passwords: A framework for comparative analysis of net authentication schemes," in Proc. IEEE Symp. Safety and privacy, 2012, pp. 553–567.

[4] A. J. Aviv, k. Gibson, E. Mossop, M. Blaze, and J. M. Smith, "Smudge attacks on smartphone contact monitors," in Proc. USENIX Workshop. Offensive Technologies, vol. 10, 2010, pp. 1–7.

[5] C. Ma, D. Wang, and S. Zhao, "Safety flaws in two elevated far-away person authentication schemes making use of wise cards," Int. J. Communique programs, vol. 27, no. 10, pp. 2215–2227, 2014.

[6] Ninuma, U. Park, and A. Ok. Jain, "soft biometric qualities for continuous consumer authentication," IEEE Trans. Information Forensics and safety, vol. 5, no. Four, pp. 771–780, 2010.

[7] Egelman, S. Jain, R.S. Portnoff, K. Liao, S. Consolvo, and D. Wagner, "Are you capable to lock?" in Proc. ACM Conf. Pc and Communications security, 2014, pp. 750–761.

[8] C. Shen, Z. Cai, and X. Guan, "continuous authentication for mouse dynamics: A sample-development technique," in Proc. IEEE Int. Conf. Responsible systems and Networks, 2012, pp. 1–12.

[9] A. Pantelopoulos and N. G. Bourbakis, "A survey on wearable sensor-based techniques for wellbeing monitoring and prognosis," IEEE Trans. Techniques, Man, and Cybernetics, vol. Forty, no. 1, pp. 1–12, 2010.

[10] R. Gravina, P. Alinia, H. Ghasemzadeh, and G. Fortino, "Multi-sensor fusion in physique sensor networks: contemporary and study challenges," Information Fusion, vol. 35, pp. 2017.

[11] Development tendencies, patron attitudes, and why smart watches will dominate," http://www.Businessinsider.Com/the-wearable-computing-marketreport-2014-10, accessed: 08-1-2015.

[12] M. Mihajlov and B. Jerman-Blažič, "On designing usable and cozy consciousness-established graphical authentication mechanisms," Interacting with desktops, vol. 23, no. 6, pp. 582–593, 2011.

[13] k. Niinuma and A. Okay. Jain, "continuous person authentication making use of temporal know-how," in Proc. SPIE protection, safety, and Sensing, 2010, p. 76670L.

[14] S. Liu and M. Silverman, "A functional advisor to biometric protection technological know-how," IEEE IT legitimate, vol. 3, no. 1, pp. 27–32, 2001.

[15] W. Shi, J. Yang, Y. Jiang, F. Yang, and Y. Xiong, "Senguard: Passive user identification on smartphones utilizing a couple of sensors," in Proc. IEEE Int. Conf. Wireless and cell Computing, Networking and Communications, 2011, pp. 141–148.

[16] Ullah, K., Shah, M.A. and Zhang, S. Effective ways to use Internet of Things in the field of medical and smart health care. International Conference on Intelligent Systems Engineering, 2016, 372–379.

[17] Nanskishor, B.R., Shinde, A. and Malathi, P. Android Based Body Area Network for the Evaluation of Medical Parameters. IEEE Intelligent Solutions in Embedded Systems (WISES), 2012.

[18] Figueredo, M.V.M. and Dias, J.S. Mobile Telemedicine System for Home Care and Patient Monitoring. Proceedings of the 26th Annual International Conference of the IEEE, 2004.

[19] Huang, H., Gong, T., Ye, N., Wang, R. and Dou, Y. Private and secured medical data transmission and analysis for wireless sensing healthcare

system. IEEE Transactions on Industrial Informatics 13 (3) (2017) 1227–1237.

[20] Sawand, A., Djahel, S., Zhang, Z. and Naït-Abdesselam, F. Toward "Energy-Efficient and Trustworthy eHealth Monitoring System". China Commun. 12 (1) (2015) 46–65.

[21] Fan, R., Ping, L.D., Fu, J.Q. and Pan, X.Z. The new secure and efficient data storage approaches for Wireless Body Area Networks. International Conference on Wireless Communications and Signal Processing (WCSP), 2010, 1–5.

[22] Wang, C., Zhang, B., Ren, K., Roveda, J.M., Chen, C.W. and Xu, Z. A Privacy-aware Cloud-assisted Healthcare Monitoring System via Compressive Sensing. Proc. of 33rd IEEE INFOCOM, 2014, 21302138.

[23] Kirtana, R.N. and Lokeswari, Y.V. An IoT Based Remote HRV Monitoring Systemfor Hypertensive Patients. IEEE International Conference on Computer, Communication, and Signal Processing, 2017.

[24] Ibraimi, L., Asim, M. and Petković, M. Secure management of personal health records by applying attribute-based encryption. 6th International Workshop on Wearable Micro and Nano Technologies for Personalized Health (pHealth), 2009, 71–74.

[25] Shruti Bhargava Choubey Adunoori Shalini, MCNN based healthcare surveillance system Using IOT, Journal of International Pharmaceutical Research, vol. 1, issue 1, 676–682, 2019.

[26] Mohammed, J., Lung, C.H., Ocneanu, A., Thakral, A., Jones, C. and Adler, A. Internet of Things: Remote patient monitoring using web services and cloud computing. IEEE International Conference on, and Green Computing and Communications (GreenCom) IEEE and Cyber, Physical and Social Computing (CPSCom), 2014, 256–263.

[27] Raghav, K., Paul, J. and Pandurangan, R. Design and Development of E-Health Care Monitoring System. International Conference on Applied Internet and Information Technologies, 2016.

[28] Oliver, N. and Flores-Mangas, F. Health Gear: a real-time wearable system for monitoring and analyzing physiological signals. International Workshop on Wearable and Implantable Body Sensor Networks, 2006.

[29] Haiping Huang, Tianhe Gong, Ning Ye, Ruchuan Wang and Yi Dou "Private and Secured Medical Data Transmission and Analysis for Wireless Sensing Healthcare System"IEEE Transactions on Industrial Informatics , Volume: 13, Issue: 3, June 2017

[30] Alii, D., Suresh, P, "An overview of research issues in the modem healthcare", American Journal of Applied Sciences, vol. 9, no. 1, pp. 54–59, 2012.

[31] Sowmyasudhan S and Manjunath S, "A wireless based real-time Patient monitoring system", International Journal of Scientific & Engineering Research, vol. 2, no. 11, Nov. 2011.

[32] Edward Teaw, Guofeng Hou, Michael Gouzman, K. Wendy Tang, Matthew Kane, Amy Kesluk and Jason Farrell, "A Wireless Health Monitoring System", International Conference on Information Acquisition, Print ISBN: 0-7803-9303-1, June 27–July 3, 2005.

[33] Zimu Li, Guodong Feng, Fenghe Liu, Jia Q Dong, RidhaKamoua and Wendy Tang, "Wireless Health Monitoring System", Applications and Technology Conference (LISAT), pp. 1–4, 2010.

[34] Isais R, Nguyen K., Perez G, Rubio R, and Nazeran H, "A low-cost microcontroller-based wireless ECG-blood pressure telemonitor for home care," Engineering in Medicine and Biology Society, proceedings of the 25th Annual International Conference of the IEEE, vol. 4, pp. 3157–3160, 2003.

[35] Priya, B., Rajendran, S., Bala, R. and Gobbi, R., "Remote Wireless Health Monitoring Systems," Innovative Technologies in Intelligent Systems and Industrial Applications- CITISLA, pp. 383–388, 2009.

[36] Jubadi, W.M. and MohdSahak, S.F.A., "Heartbeat Monitoring Alert via SMS ", IEEE Symposium on Industrial Electronics &ApplicationsISIEA, pp. 1–5, 2009.

[37] AIMejrad, A.S., "A Single Supply Standard 805 1 Microcontroller based Medical K-grade Isolation ECG Module with Graphics LCD," Second International Conference on Intelligent System Design and Engineering Application -ISDEA, pp. 1184–1187, 2012.

[38] MoeenHassanalieragh, Alex Page, Tolga Soyata, Gaurav Sharma, Mehmet Aktas, Gonzalo Mateos, BurakKantarci, Silvana Andreescu, Health Monitoring and Management Using Internet-of-Things (IoT) Sensing with Cloud-Based Processing: Opportunities and Challenges, 2015.

[39] M. Shamim Ghulam Muhammad, Cloud-assisted Industrial Internet of Things (iiot) - Enabled framework for health monitoring, 2016.

[40] H. S. Park, H. M. Lee, HojjatAdeli, I. Lee, A New Approach for Health Monitoring of Structures: Terrestrial Laser Scanning, 2006.

[41] Nicola Bui, Michele Zorzi, Health care applications: a solution based on the internet of things, 2011.

[42] Min Chen, Yujun Ma, Jeungeun Song, Chin-Feng Lai, Bin Hu, Smart Clothing: Connecting Human with Clouds and Big Data for Sustainable Health Monitoring, 2016.

[43] Shruti Bhargava Choubey, Adunoori Shalini, A Survey on Real Time Health Care Monitoring System, Jour of Adv Research in Dynamical & Control Systems, vol. 10, 2018.

[44] Hong yang Zhang, Junqi Guo, Xiaobo Xie, RongfangBie, Yunchua-nun, Environmental Effect Removal Based Structural Health Monitoring in the Internet of Things, 2013.

[45] CharithPerera, Arkady Zaslavsky, Peter Christen, DimitriosGeor-gakopoulos, Sensing as a service model for smart cities supported by Internet of Things, 2013.

[46] Lih, O.S.; Jahmunah, V.; San, T.R.; Ciaccio, E.J.; Yamakawa, T.; Tanabe, M.; Kobayashi, M.; Faust, O.; Acharya, U.R. Comprehensive electrocardiographic diagnosis based on deep learning. Artif. Intell. Med. 2020, 103, 101789

[47] Yin, M., Tang, R. Liu, M., Han, K., Lv, X., Huang, M., Xu, P. Hu, Y. Ma, B. Gai, Y., "Influence of Optimization Design Based on Artificial Intel-ligence and Internet of Things on the Electrocardiogram Monitoring System" J. Healthc. Eng. 2020, 2020, 8840910.

[48] Huda, N.; Khan, S.; Abid, R.; Shuvo, S.B.; Labib, M.M.; Hasan, T. A Low-cost, Low-energy Wearable ECG System with Cloud-Based Arrhythmia Detection. In Proceedings of the 2020 IEEE Region 10 Symposium (TENSYMP), Dhaka, Bangladesh, 5–7 June 2020; pp. 1840–1843

[49] AD Acharya and SN Patil, "IOT based Health Care Monitoring", *IEEE*, pp. 363-8, 2020.

[50] Saghiri, A. M., HamlAbadi, K. G., & Vahdati, M. The internet of things, artificial intelligence, and blockchain: implementation perspectives. In S. Kim & G. C. Deka (Eds.), Advanced applications of blockchain technology Springer (pp. 15–54), 2020.

Index

About the Editors

M. A. Jabbar is a Professor and Head of the Department AI&ML, Vardhaman College of Engineering, Hyderabad, Telangana, India. He obtained Doctor of Philosophy (Ph.D.) in the year 2015 from JNTUH, Hyderabad, and Telangana, India. He has been teaching for more than 20 years. His research interests include Artificial Intelligence, Big Data Analytics, Bio-Informatics, Cyber Security, Machine Learning, Attack Graphs, and Intrusion Detection Systems.

Kantipudi MVV Prasad received his B.Tech degree in Electronics & communications engineering from ASR College of Engineering, Tanuku, India, the M.Tech degree in Digital Electronics and Communication Systems from Godavari Institute of Engineering & Technology, Rajahmundry, India, Ph.D. from BITS, VTU, Belgum and currently working as Director Of Advancements, Sreyas Institute Of Engineering & Technology, Hyderabad, having teaching experience around 10 years. Previously he was working with RK University, Rajkot. His current research interests are in Signal Processing and Machine Learning, Education and Research.

Sheng-Lung Peng obtained Doctor of Philosophy in Computer Science, National Tsing Hua University, Taiwan, 1999. Thesis: A study of graph searching on special graphs. He is a Professor, Department of Computer Science and Information Engineering, National Dong Hwa University, Hualien, Taiwan. Presently he is a Director, Information Service Association of Chinese Colleges, Taiwan and Taiwan Regional Contest Director of ACM–ICPC.

Mamun Bin Ibne Reaz was born in Bangladesh, in December 1963. He received his B.Sc. and M.Sc. degree in Applied Physics and Electronics, both from University of Rajhashi, Bangladesh, in 1985 and 1986, respectively. He received his D.Eng. degree in 2007 from Ibaraki University, Japan. He is currently an Associate Professor in the Universiti Kebangsaan Malaysia,

Malaysia involving in teaching, research and industrial consultation. He is a regular associate of the Abdus Salam International Center for Theoretical Physics since 2008. He has vast research experiences in Norway, Ireland and Malaysia. He has published extensively in the area of IC Design and Biomedical application IC. He is author and co-author of more than 100 research articles in design automation and IC design for biomedical applications.

Ana Madureira was born in Moçambique, in 1969. She got her BSc degree in Computer Engineering in 1993 from ISEP, Master degree in Electrical and Computers Engineering Industrial Informatics, in 1996, from FEUP, and the PhD degree in Production and Systems, in 2003, from University of Minho, Portugal. She became IEEE Senior Member in 2010. She had been Chair of IEEE Portugal Section (2015–2017), Vice-chair of IEEE Portugal Section. Teacher Coordinator (Polytechnic Teacher) Instituto Politécnico do Porto Instituto Superior de Engenharia do Porto, Portugal.